Jerry Carr-Brion
Process Development

Also of Interest

Process Engineering.
Addressing the Gap between Study and Chemical Industry
Michael Kleiber, 2020
ISBN 978-3-11-065764-7, e-ISBN (PDF) 978-3-11-065768-5,
e-ISBN (EPUB) 978-3-11-065807-1

Process Machinery.
Commissioning and Startup – An Essential Asset Management
Activity
Fred K. Geitner, Ronald G. Eierman, 2021
ISBN 978-3-11-070097-8, e-ISBN (PDF) 978-3-11-070107-4,
e-ISBN (EPUB) 978-3-11-070113-5

Technoscientific Research.
Methodological and Ethical Aspects
Roman Z. Morawski, 2019
ISBN 978-3-11-058390-8, e-ISBN (PDF) 978-3-11-058406-6,
e-ISBN (EPUB) 978-3-11-058412-7

Sustainable Process Engineering
Gyorgy Szekely, 2021
ISBN 978-3-11-071712-9, e-ISBN (PDF) 978-3-11-071713-6,
e-ISBN (EPUB) 978-3-11-071730-3

Fluid Machinery.
Life Extension of Pumps, Gas Compressors and Drivers
Heinz Bloch, 2020
ISBN 978-3-11-067413-2, e-ISBN (PDF) 978-3-11-067415-6,
e-ISBN (EPUB) 978-3-11-067427-9

Jerry Carr-Brion

Process Development

—

An Introduction for Chemists

DE GRUYTER

Author
Dr. Jerry Carr-Brion
Histon
Cambridge
United Kingdom

ISBN 978-3-11-071786-0
e-ISBN (PDF) 978-3-11-071787-7
e-ISBN (EPUB) 978-3-11-071791-4

Library of Congress Control Number: 2021951714

Bibliographic information published by the Deutsche Nationalbibliothek
The Deutsche Nationalbibliothek lists this publication in the Deutsche Nationalbibliografie;
detailed bibliographic data are available on the Internet at http://dnb.dnb.de.

Contents

Introduction

Few disciplines are as fascinating and rewarding as that of chemical process development. At times it can also seem like one of the most frustrating occupations, but that is in its nature. The highs would not be as high were it not for the lows. A good process development chemist has to straddle many disciplines. In addition to a sound knowledge of synthetic chemistry, he or she will need to learn something about chemical engineering, statistics, analytical chemistry, costing, patent law, quality assurance and hazard evaluation. There is always something new, for process development does not stand still. One project will follow another and new techniques and regulatory requirements appear all the time. However, those that stay the course will eventually have the satisfaction of being able to point to a pharmaceutical, fragrance, industrial chemical or agrochemical product and say, 'Our process made that'.

Process development is a team effort. That is not to say that individual flashes of genius cannot make a huge difference, but today's industry requires diverse teams of people with different skills and responsibilities working cooperatively towards a goal. Once a project is successfully carried out and the product from the new process goes into commercial production, it is amazing how many people (some more deservedly than others) feel their contribution was the one that mattered. It has been said that success has a thousand fathers, but failure is an orphan – that is the way of the world.

This book will concentrate on process development in the pharmaceutical, agrochemical and fragrance/flavour industries. However, many of the techniques described are equally applicable to bulk chemical manufacturing. Of course, some pharmaceuticals and agrochemicals are made on a very large scale, and so could be considered 'bulk chemicals', but there are clear differences in the regulatory regimen applied. Paracetamol (acetaminophen) is an example of a simple pharmaceutical compound produced on a huge scale, but the regulations and analysis required for it are quite different to that used for other non-pharmaceutical bulk aromatics.

Typically, a process development project starts with a compound previously made on a small scale in a research laboratory. Thus, a few grams of material may have been made on a number of occasions using small laboratory flasks. The synthetic route used to make the compound may well have been chosen as the most convenient route to make hundreds or thousands of analogues (indeed, the initial chemistry may have been chosen because it is the most amenable to synthesis by robotic systems). However, the process development chemist will be lucky indeed if the initial route is the cheapest and most amenable to scale up. Tempting as it is to persevere with what is 'known to work', this is often a costly mistake. There is a great saving in time and money to switch to the 'best' route sooner rather than later. However, since very tight deadlines for material delivery are often given, the development chemist

https://doi.org/10.1515/9783110717877-001

may have little choice but to make the first development batches by a research route that is unlikely to be the final route of choice. The scouting of alternative routes and reagents to make a compound is often referred to as process research, and larger companies may have separate process research groups distinct from process development groups. Generally, however, the same team that looks at different routes will aim to carry the final process through until a production process is reached.

Process development projects may also start with a compound that has been in production for many years. This is the case with generic pharmaceuticals or agrochemicals. Here, the constraints may be existing patents: the patent on the compound may have expired, but process patents, intermediate patents, polymorph patents, etc. can still be in force. The route used in the original patent for the compound may not be easy to scale up, or may be more costly than other routes. Patent holders will be out to make your life as difficult as possible, so sometimes novel synthetic routes are required even with long-established compounds.

Once a route is found, typically the cheapest route that looks capable of delivering a high-purity product, then optimisation and scale-up can begin in earnest. The optimisation of conditions is often carried out using statistical methods such as experimental design (D of E). Automation has made optimisation less tedious and less prone to error. Variables such as temperature profile, pH and addition time of reagents can be tightly controlled and optimised. The chemist's skill comes in knowing which variables are likely to be important and what values of variables are likely to give sensible outcomes. These issues will be discussed in depth in later chapters.

As a process is scaled up, the hazards involved become more significant. A tiny exotherm on a small-scale reaction may be quite capable of turning a plant into a heap of twisted metal when a large batch is prepared. Safety evaluation is an essential part of process development. Sometimes, development chemists are criticised for being too cautious, but better be cautious than dead.

Processes, typically, start as small-scale laboratory processes – ultimately, the goal is to run them on a full-scale plant. This raises the question as to whether a particular reaction should be run on an intermediate scale. A plant can be mimicked, to some extent, using relatively small-scale laboratory equipment, such as jacketed vessels and small pumps. Sometimes this is referred to as a mini-plant. Data on exotherms, separations and so on can be gathered at this point, and the reaction optimised. However, generally, the reaction will then be carried out on large-scale glass equipment (the facility where this occurs may be referred to as a kilo-lab, although various other names are used in different companies) and/or in a pilot plant containing smaller versions of plant vessels and equipment. It is a debatable point as to whether it is desirable to go straight from small-scale laboratory vessels to the full-scale plant. Most chemists prefer to see a reaction working at a pilot plant scale, regardless of the quantity of laboratory data that has been gathered, before scaling up to a full-scale production plant. There is always a possibility that something has been missed. In practice, kilo-lab and pilot plant batches are often needed to provide material for toxicology testing, clinical trials, field trials, etc., so in these cases the process chemist gets to look at reactions on an intermediate scale before full plant production is required.

Last, but not least, process chemists need to have environmental considerations in the forefront of their minds. Processes that initially involve chlorinated solvents or heavy metals may well have to be replaced with more benign alternatives. It should be noted that it is not enough to simply remove heavy metals from a product; they have to be demonstrably removed, with all the regulatory and analytical costs involved – far better not to use them in the first place. A related consideration in the pharmaceutical field is the control of potentially mutagenic compounds. Again, there are significant regulatory and analytical costs involved in demonstrating their removal – far better to avoid their formation.

Learning is an ongoing process for any professional process development chemist. It is worthwhile attending some of the excellent training courses, webinars and conferences run by the Scientific Update organisation in a number of countries. A useful annual process development symposium is run by the Society for Chemical Industry (SCI) in Cambridge, UK.

Finally, I would like to thank all my former colleagues at PPF/Quest, Schering/ AgrEvo/Aventis Cropscience and Resolution Chemicals for putting up with me over the years. In particular, I would like to thank Parveen Bhatarah and Mike Pollard for their invaluable help with the preparation of this book.

Chapter 1
Route, reagent and solvent selection

1.1 Basic route selection

There are always a huge number of ways in which you could conceivably make any reasonably complex molecule, and the methods that have been used previously are not necessarily the most efficient. Certain guidelines are useful in planning a synthesis. Firstly, we can consider the two synthetic routes below, both forming the final product D:

$$A \to B \to C \to D$$

and

$$E \to F; \quad G \to H; \quad F + H \to D$$

Both reaction schemes consist of three reactions. If each step has a molar yield of 90%, the top linear reaction scheme would give an overall molar yield of 72.9%, while the lower convergent synthesis would give an overall molar yield of 81.0%. Usually, it is best to disconnect larger molecules into two or three roughly equal portions in order to maximise yields, rather than continually extend a starting material a little at a time.

All syntheses must start from commercially available starting materials, the cheaper the better. A knowledge of which molecules are relatively inexpensive is very useful. For chiral products, it is often beneficial to select starting materials that have the correct chirality 'built in' – the so-called chiral pool.

The obvious basic building blocks are the starting materials for many syntheses, but other starting materials are less evident. For example, furfural (Fig. 1.1), which is isolated from farm waste such as corn cobs and oat husks, is a relatively inexpensive starting material.

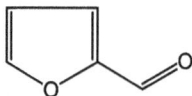

Fig. 1.1: Furfural, a useful starting material.

However, even the putting together of 'simple' building blocks may not always be straightforward. Consider the formation of a 'mesityl oxide portion' of a larger molecule, which might be required as an intermediate for a pharmaceutical or fragrance compound. Here the obvious synthetic method (Fig. 1.2) is the aldol condensation of a methyl ketone (**a**) with acetone, to give the hydroxyketone (**b**), which readily dehydrates with acid to give the mesityl oxide analogue (**c**). However, simply reacting (**a**) with acetone in the presence of a base gives a complex mixture of products, since both compound (**a**) and acetone can react with themselves and also react with

https://doi.org/10.1515/9783110717877-002

each other in more than one way, giving large quantities of various unwanted condensation products. The literature [1] suggests the reaction is carried out using lithium diisopropylamide (LDA) as base at low temperatures, giving the kinetic enolate
of (a), and then slowly adding acetone. Such a procedure gives good yields of (b),
but is very expensive to implement on a plant scale, where temperatures below
those obtained with chilled ethylene glycol (around −15 to −20 °C in practice) are
difficult to reach in standard vessels. Also, LDA, though available in solution in
bulk metal drums, is a relatively expensive reagent to use.

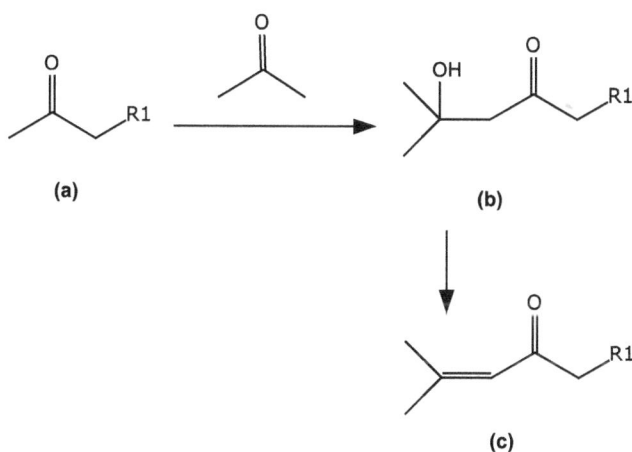

Fig. 1.2: Aldol condensation of a methyl ketone with acetone.

A totally different approach would be a Friedel-Crafts acylation (Fig. 1.3) between the
acid chloride (d) and isobutylene in the presence of a Lewis acid, to give a mixture of
(c) and the chloride (e), which readily loses HCl in the presence of a base to give (c).
Isobutylene has a boiling point of about −7 °C and is a useful reactant in Friedel-
Crafts reactions with acid chlorides [2], which can be carried out at −10 to −20 °C.
Isobutylene does form dimers to some extent, but these octene compounds are readily distilled out, unlike acetone condensation products. Such reactions might require
a stoichiometric amount of a Lewis acid (TiCl$_4$ or AlCl$_3$ are likely candidates) and
would require a solvent that is compatible with such acids. Dichloromethane is an
obvious choice, but chlorinated solvents are frowned upon nowadays for environmental reasons.

Other routes to (c) involving metal coupling reactions or via various enol compounds are also possible, but the point is that even a 'simple' transformation may
not be straightforward on a plant scale, and careful consideration of the likely capital, raw material and running costs are needed, along with an examination of the
environmental and safety factors, in order to choose the best route for the process.

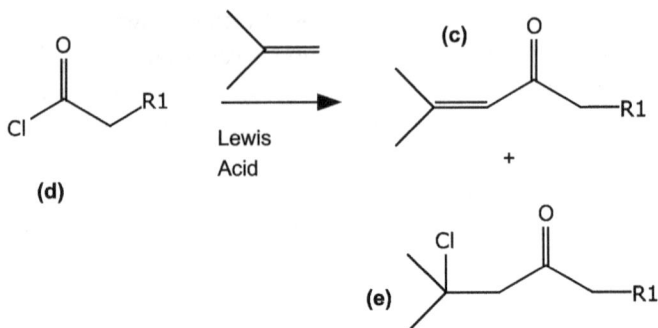

Fig. 1.3: Acylation of isobutylene.

1.2 Syntheses from chiral starting materials

A number of naturally occurring terpenoid compounds are inexpensive, and some act as a useful source of chirality. For example, (R)-limonene (Fig. 1.4) is readily available from orange peel, etc.

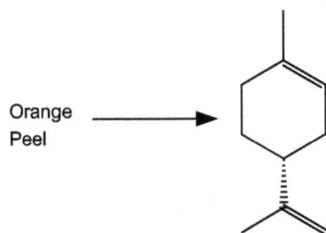

Fig. 1.4: Limonene, a useful chiral starting material.

(R)-Limonene (Fig. 1.5, **f**) has been used in cannabinoid syntheses. Conversion to the key intermediate p-mentha-2,8-dien-1-ol (**g**) was initially accomplished using a rose-bengal photosensitiser to generate singlet oxygen [3], but the use of the latter species is challenging on a large scale. However, a recent paper [4] described a version of this reaction using tetraphenylporphyrin, rather than rose bengal, as photosensitiser in a continuous reactor. A selectivity of 66% was achieved with 55% conversion and an output of around 66 g/day. Further development might lead to a commercial process.

A longer synthesis (Fig. 1.6) was devised by Firmenich chemists [5], avoiding the use of singlet oxygen. Initial oxidation of (R)-limonene (**f**) gave the epoxide (**h**), which was selectively reacted with thiophenol to give the sulfide (**i**). Hydrogen peroxide gave the sulfoxide (**j**), which underwent elimination to (**g**) on heating at 425 °C.

Reaction of the intermediate p-mentha-2,8-dien-1-ol (Fig. 1.7, **g**) with olivetol (**k**) gives a mixture of different cannabinoids, the predominant species present depending

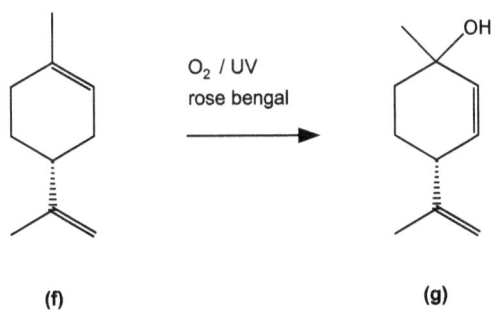

Fig. 1.5: Preparation of *p*-mentha-2,8-dien-1-ol (**g**) using singlet oxygen.

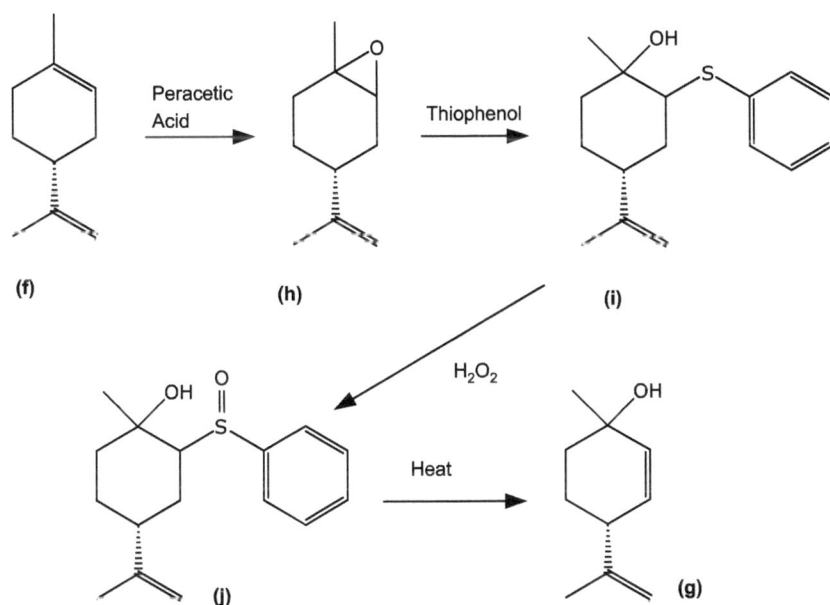

Fig. 1.6: Preparation of *p*-mentha-2,8-dien-1-ol (**g**) via sulfoxide elimination.

on the conditions [3] employed. Mild acids give mainly cannabidiol (CBD, **l**), stronger acids, such as p-toluene sulfonic acid (p-TSA), give mainly Δ8-THC (**m**), while reacting with boron trifluoride and magnesium sulphate in dichloromethane at 0 °C gives mainly Δ9-THC (**n**), the psychoactive constituent of smoked marijuana.

A related scheme of reactions (Fig. 1.8), producing CBD (**l**) first from (**g**) and (**k**) (or related compounds) with metal triflate catalysts, followed (if required) by the conversion of CBD (**l**) to Δ9-THC (**n**) with aluminium alkyls, was described in a patent [6] from Albany Research.

Other terpenoid starting materials are more obscure than (*R*)-limonene. For example, sclareol (Fig. 1.9, **o**), isolated from the clary sage plant (*Salvia sclarea*), is used to

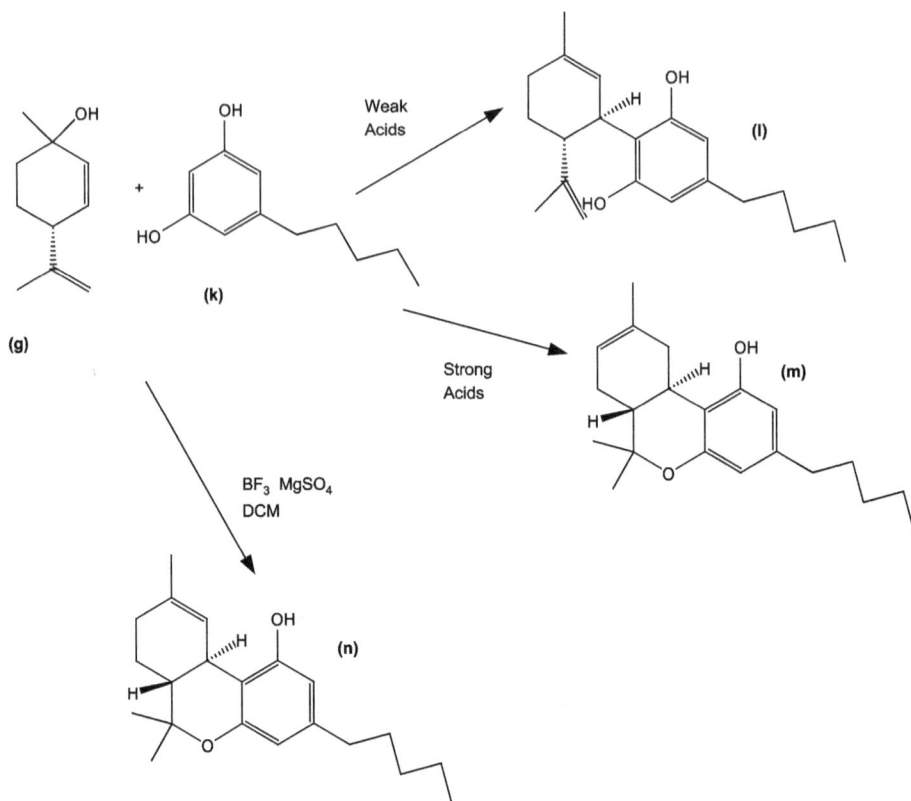

Fig. 1.7: Conversion of p-mentha-2,8-dien-1-ol (**g**) to cannabinoids using acids.

make the important amber fragrance compound Ambrox® (**p**) (ambroxide); a number of routes are reported in the literature [7, 8]. A good knowledge of the various branches of natural product chemistry can sometimes be helpful in devising new synthetic pathways.

Some fairly complex molecules are cheap since they are produced in bulk for another use entirely, but can be used as starting materials. An example is ascorbic acid (vitamin C) (Fig. 1.10), which is a convenient source of chirality.

Various sugars, amino-acids and compounds derived from them are useful starting materials [9] for chiral organic syntheses, and their use can eliminate difficult resolution steps.

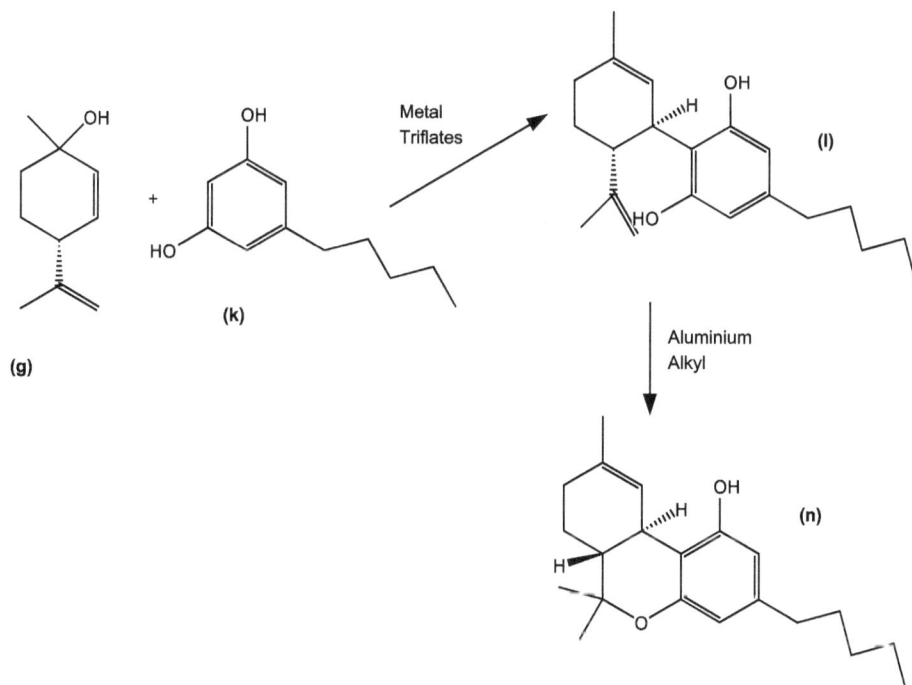

Fig. 1.8: Conversion of *p*-mentha-2,8-dien-1-ol (**g**) to CBD (**l**) with metal triflate.

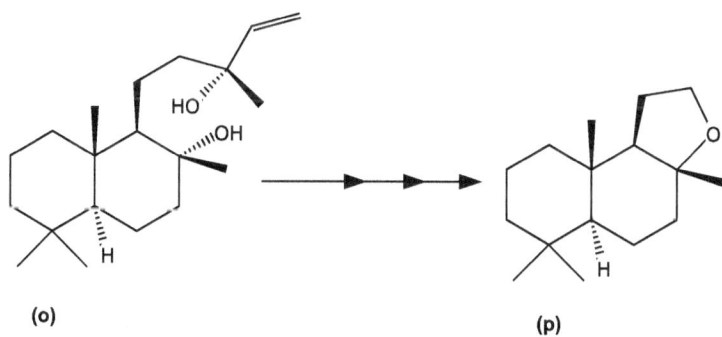

Fig. 1.9: Ambrox® (**p**) is made from sclareol (**o**).

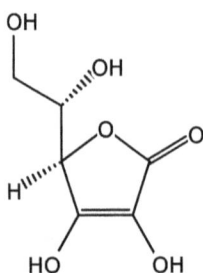

Fig. 1.10: Ascorbic acid (vitamin C), another useful chiral starting material.

1.3 Examples of convergent syntheses (prostaglandins)

An interesting example of convergent organic synthesis is a Scinopharm patent [10] detailing the synthesis of the synthetic prostaglandin travoprost (Fig. 1.11), which is used to treat glaucoma. The same patent describes an analogous synthesis of the related compound bimatoprost.

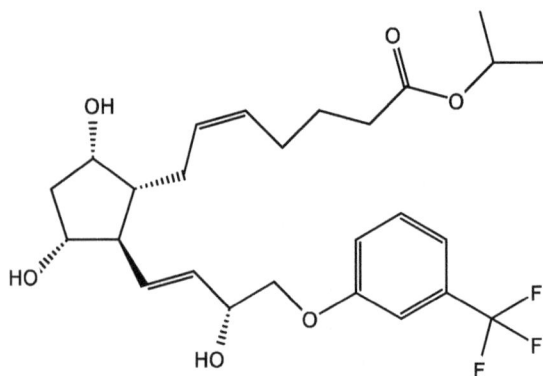

Fig. 1.11: Travoprost, a prostaglandin used to treat glaucoma.

These prostaglandin compounds are extremely active and are used in doses of a few micrograms. Chromatographic separations are also required to give adequate purity, since the final products and most of the intermediates are oils that do not crystallise. Production batch sizes therefore tend to be unusually small, typically only 1 or 2 kg of the final product. The key step in the synthesis (Fig. 1.12) is the Michael addition of a lithium cuprate compound to the cyclic enone (**q**), to give the protected prostaglandin structure (**r**). The cuprate (presumably unstable) is formed from the tin derivative (**s**) in situ.

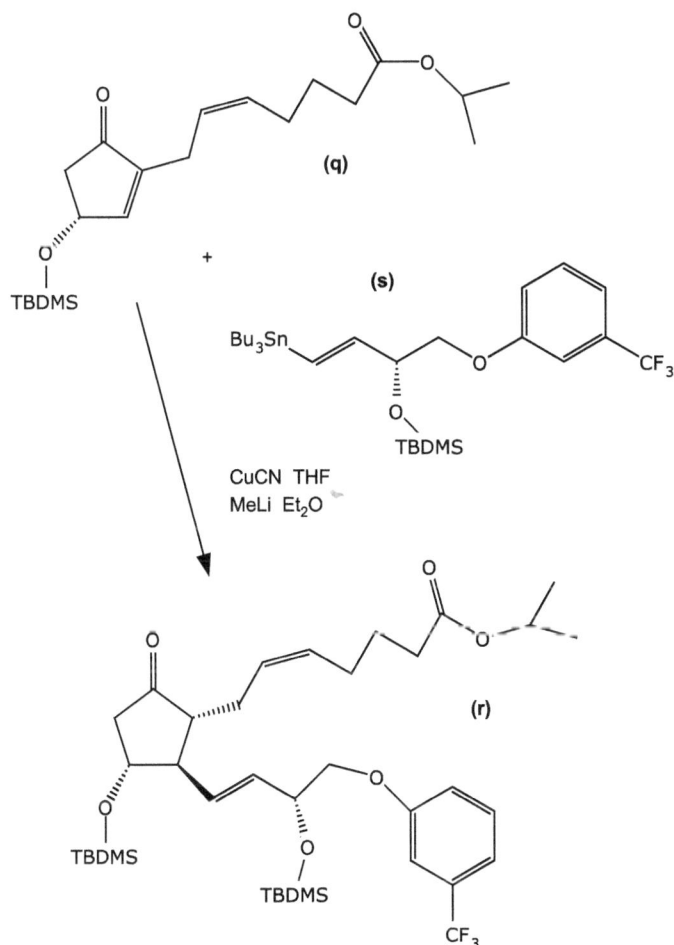

Fig. 1.12: Scinopharm prostaglandin synthesis (Michael addition).

The starting material for the five-membered ring is our old friend furfural, which is elaborated to give the intermediate (Fig. 1.13, **t**). The latter compound is rearranged with zinc chloride to give a mixture of the required ketone intermediate (**u**) and the unwanted isomer (**v**).

Separation of (**u**) and (**v**) was carried out by column chromatography on the chloral adducts. Subsequent enzymatic resolution gave the required enantiomer of (**u**). Although this route is a good example of convergent synthesis, the use of tin (a heavy metal with adverse environmental effects) is an issue that should be addressed if the process is to be used for production.

Traditionally, prostaglandins such as bimatoprost, latanoprost and travoprost are produced from the so-called 'Corey lactone' (Fig. 1.14, **w**), where R1 is a protecting

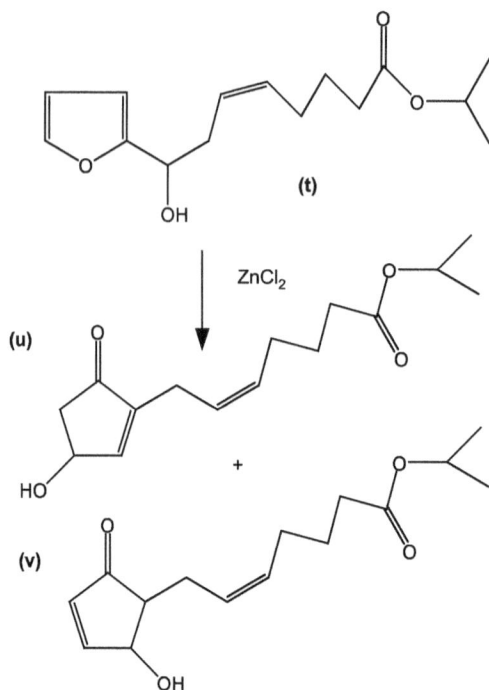

Fig. 1.13: Scinopharm prostaglandin synthesis (rearrangement).

group, typically benzoyl or 4-phenylbenzoyl [11, 12, 13]. The lactone is commercially available, but expensive. The *trans* double bond can be added using a Horner–Wadsworth–Emmons (HWE) phosphonate reaction (after the oxidation of the free alcohol to an aldehyde, **x**) – addition of the phosphonate (**y**) gives a lactone (**z**), containing the 'lower' chain.

The *cis* double bond is formed later in the synthesis (Fig. 1.15) using a conventional Wittig reaction, in which the lactol compound (**aa**) (protected at this point with silyl or tetrahydropyranyl groups, R2) is reacted with the phosphonium salt (**ab**) to give a precursor (**ac**) to the final product. In contrast to the HWE reaction, which strongly favours the *trans* double bond, the Wittig reaction of an unstabilised ylide can be optimised to give the *cis* double bond with very little *trans* isomer.

Thus, the conventional prostaglandin synthesis uses three portions, the central Corey lactone, the upper chain and the lower chain. One interesting variant from Chirotech [14] reacts the tricyclic compound (Fig. 1.16 **ad**) with the cuprate (**ae**), which provides the lower chain, to give the ketone (**af**). Subsequent Baeyer–Villiger rearrangement gave the crystalline lactone (**ag**). This is analogous to the lactone (**z**) used in the traditional synthesis, differing only in the OH group forming the lactone, and can be carried on to the final product in a similar manner.

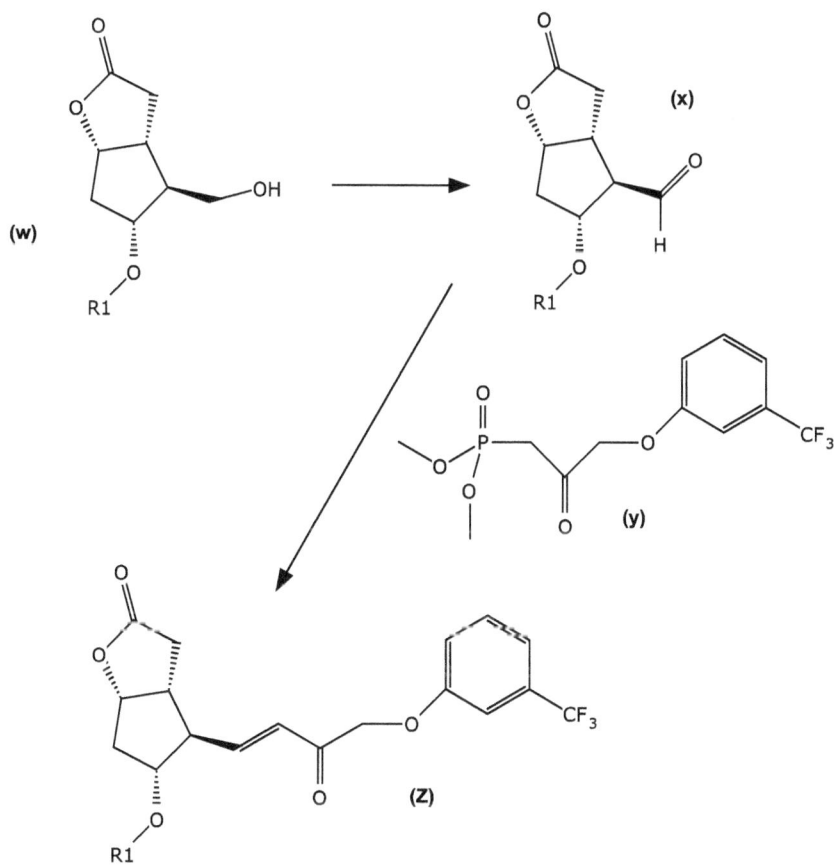

Fig. 1.14: Traditional prostaglandin synthesis (lower chain addition).

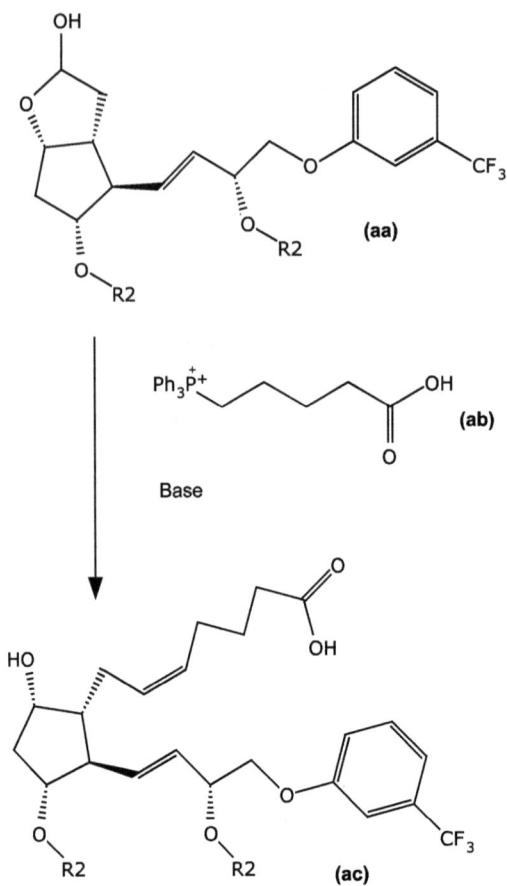

Fig. 1.15: Traditional prostaglandin synthesis (upper chain addition).

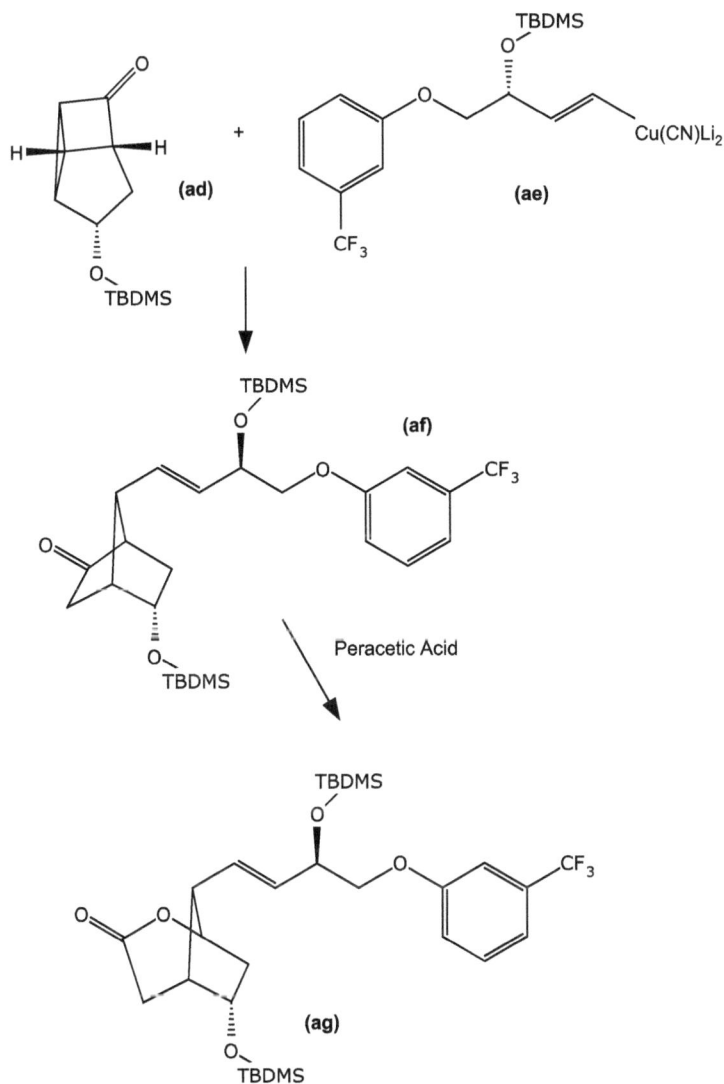

Fig. 1.16: Chirotech prostaglandin synthesis.

1.4 Rearrangements in synthesis

GSK's PPAR-α agonist GW641597X (Fig. 1.17, **ah**) is another example of convergent synthesis – the central phenyl ether being produced from the phenol (**ai**) and the chloride (**aj**) by a simple substitution in the presence of sodium hydroxide [15].

   The phenol compound (**ai**) could in theory be produced by the alkylation of 2-methylhydroquinone with a compound such as ethyl bromoisobutyrate. However,

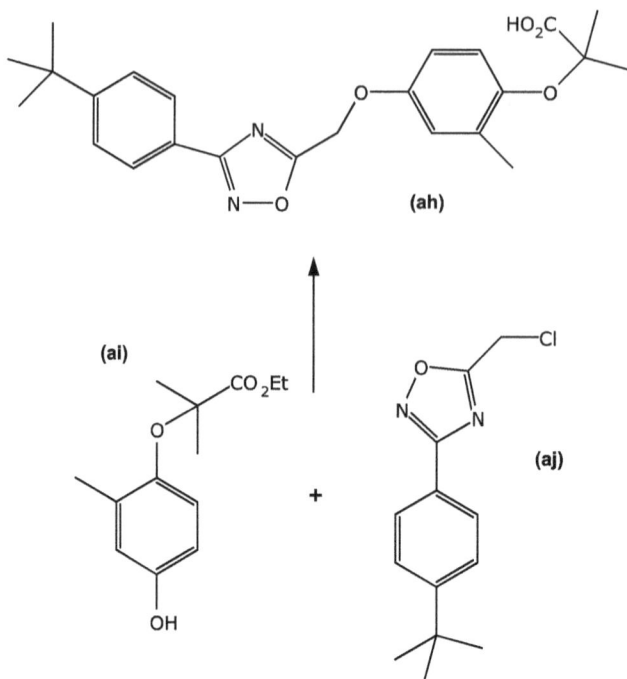

Fig. 1.17: GW641597X synthesis, phenyl ether formation.

in practice, both phenol groups would react with an alkylating agent, and any protecting group would be unlikely to be installed in a selective enough manner to be useful. However, since many substituted acetophenone compounds are readily available, it is much easier to start with the compound (Fig. 1.18, **ak**), alkylate the phenol to give the ether (**al**) and then employ a Baeyer–Villiger rearrangement to make the phenyl acetate (**am**).

Initially, the key Baeyer–Villiger rearrangement was carried out using 3-chloroperoxybenzoic acid (MCPBA). However, this peracid is known to be relatively unstable, and there are restrictions on the transport of large quantities in many countries. In addition, MCPBA has poor 'atom-economy', with only the oxygen atom being incorporated into the product out of the 16 atoms in the reagent. Sodium perborate is a more stable and cheaper oxidising agent. If acetic acid is used as a solvent, the perborate forms peracetic acid in situ – the latter compound carrying out the rearrangement. Peracetic acid is unstable, but can be safely used when produced in such a manner. A nitrogen purge was employed to remove any oxygen formed.

In general, it is important to be able to visualise rearrangement reactions as well as obvious disconnections when planning syntheses. Studying examples of rearrangements can help develop an ability to spot such transformations, just as a

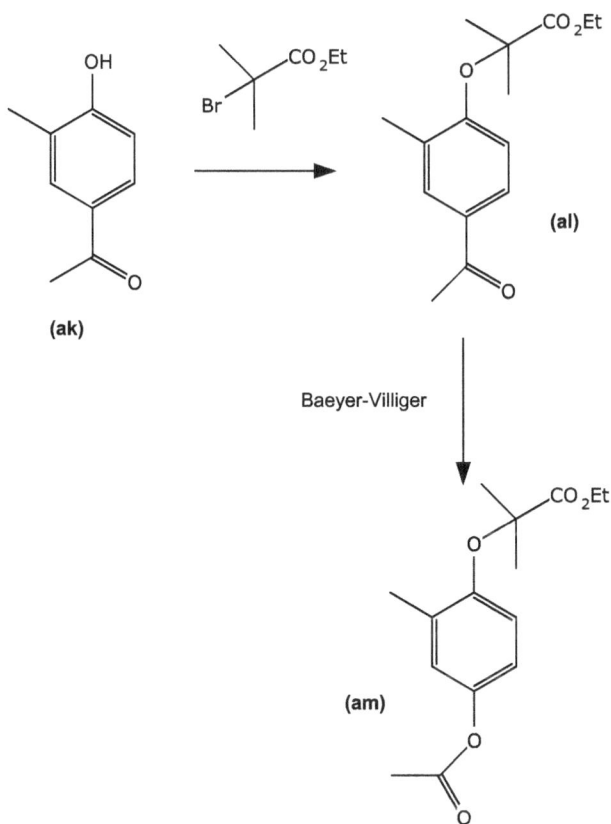

Fig. 1.18: GW641597X synthesis, Baeyer–Villiger rearrangement.

chess player learns to spot advantageous sacrifices of pieces or a bridge player opportunities for unusual bids.

Another useful rearrangement involves the synthesis of 2,6-difluorinated anilines and related compounds. The fluorination of aniline is unselective, as is the nitration of polyfluorinated aromatics. However, compounds such as 2,6-difluorobenzonitrile and 2,3,5,6-tetrafluorobenzonitrile are available and can be converted to the anilines by initial hydrolysis to the amide, followed by a Hofmann rearrangement. A Lanxess patent [16] describes the Hofmann rearrangement of 2,3,5,6-tetrafluorobenzamide (Fig. 1.19, **an**) to 2,3,5,6-tetrafluoroaniline (**ao**) using aqueous sodium hydroxide and bleach. The authors note that chlorine or bromine could be used in conjunction with sodium hydroxide, forming the hypohalite reactant in-situ.

Similarly, an older Hoechst patent [17] describes the conversion of 2-benzyloxy-6-fluorobenzamide (Fig. 1.20, **ap**) to 2-benzyloxy-6-fluoroaniline (**aq**) using sodium hydroxide and chlorine gas in the presence of excess benzyl alcohol (the benzyl alcohol presumably limits side reactions by reacting with the excess chlorine and may

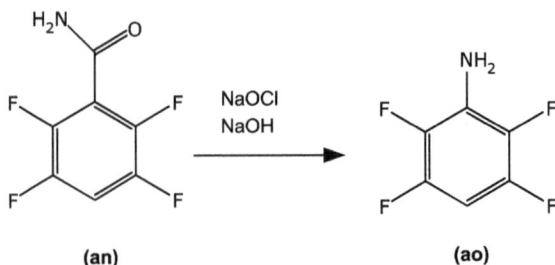

Fig. 1.19: Hofmann rearrangement (Lanxess patent).

also limit cleavage of the benzyl group). Catalytic hydrogenation of the product gave 3-fluoro-2-aminophenol (**ar**), which has been used as an intermediate for various agrochemical and pharmaceutical compounds.

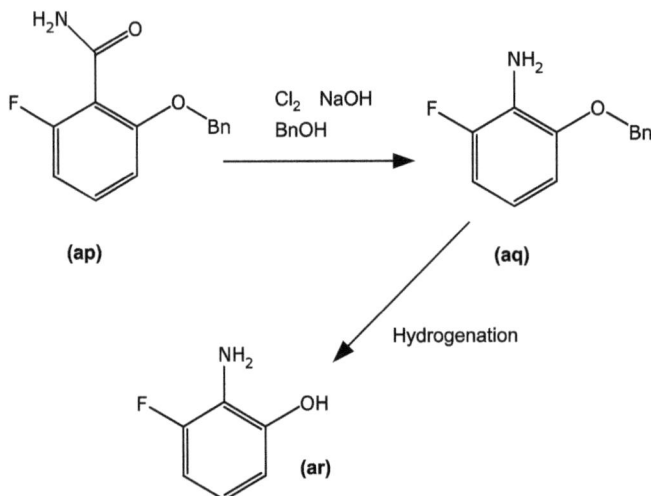

Fig. 1.20: Hofmann rearrangement (Hoechst patent).

Sigmatropic rearrangements, such as the Cope and Claisen rearrangements, are commonplace in the synthesis of fragrance compounds. For example, the key intermediate, 6-methyl-5-hepten-2-one (sulcatone) (Fig. 1.21, **as**), can be made by the Claisen rearrangement of the enol ether (**at**) [18]. In general, the Claisen and its many variants are a quick method of accessing carbonyl compounds with unsaturation at the gamma-delta bond. Further elaboration, such as metathesis, hydrobromination or epoxidation, can give rise to many useful intermediates, not solely confined to fragrances.

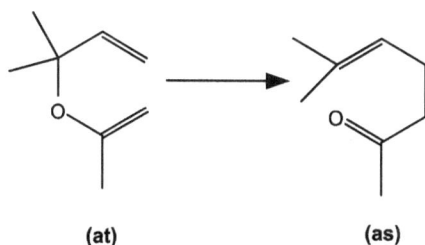

(at)   (as)

Fig. 1.21: Formation of 6-methyl-5-hepten-2-one via a Claisen rearrangement.

1.5 Listing all possible routes – brainstorming or programs?

For any target compound, the team has to ask themselves whether they have thought of every useful synthetic route. Traditionally, this involves a literature search, followed by a 'brainstorming' session with a whiteboard or flipchart (with biscuits, if you are lucky). However, nowadays computer synthetic software has advanced to a point where it can provide useful inputs to the process. Just as games such as chess and Go have succumbed to computing power, so the ability of software programs to devise practical synthetic routes can only increase. At the time of writing, the Chematica/Synthia™ program [19, 20] has shown a significant ability to devise synthetic routes, but the field is rapidly progressing, and it may be that other programs will be the market leaders in the future.

Since there are a huge number of conceivable pathways for all but the simplest of molecules, modern synthesis programs are able to highlight routes that are likely to give higher yields. For example, they can present convergent syntheses rather than the unrealistic linear routes often produced by older programs. Likely costs can also be included, so the program can highlight the more cost-effective routes. If a particularly expensive reagent has to be used, but can be employed either near the beginning or towards the end of a multi-stage synthesis, it is usually best to use it later, where less is required. Such considerations can be accounted for, so the program is able to avoid coming out with a large number of routes that will never be competitive.

1.6 Example of reagent selection – stereoselective reduction of ketones

Once a route is decided upon, reagent selection is required. Ease of handling, safety profile, environmental effects, cost and product yield are all important considerations. In some cases, supported reagents or catalysts on polymeric resins, clays, silica, charcoal or MWCNTs (multi-walled carbon nanotubes) can give better yields or purities than conventional reagents [21, 22, 23]. Their ease of filtration can make work-up and potential regeneration much simpler. One word of warning: silica gel can cause too much abrasion to glass-lined reactors, so diatomaceous silicas, such as Celite® and Celatom®, are preferred for plant use.

The stereoselective reduction of ketones to alcohols is an interesting example of a transformation where a number of different reagents can be used. BINAP metal diamine complexes (Noyori catalysts, Fig. 1.22 and analogous compounds) can selectively catalyse [24] the chiral hydrogenation of a variety of organic ketones and have been used for industrial synthesis of some pharma compounds. Both R and S catalysts are available, but expensive, so industrial use requires good conversion at low catalyst levels combined with successful recycling. The recovery of the catalyst is important both from the economic point of view and in order to minimise the contamination of the product with traces of ruthenium. Variants on standard Noyori conditions worth investigating include the use of diguanidinium- or PEG-modified BINAP catalysts, which give high e.e. values in polar solvents such as methanol and ethylene glycol [25]. The catalysts can be easily precipitated by the addition of a suitable non-polar solvent. Another potentially useful modification is Ru complexes constructed with BINAP on silica gel [26], where the catalyst is easily filtered off for re-use. However, consideration has to be given to the potential abrasive effects of silica gel on glass-lined reactors.

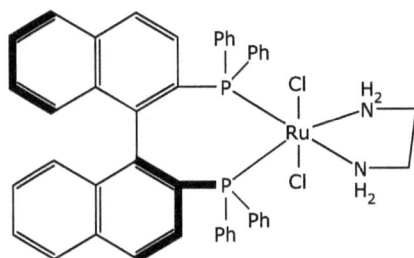

Fig. 1.22: Noyori catalyst for stereoselective hydrogenation of ketones.

Due to the expense, scarcity and toxicity of heavy metals, much research effort has been expended in devising iron complexes that can carry out chiral reductions of ketones [27, 28]. Although high e.e. (enantiomeric excess) values are far from general, they can be obtained in some cases, such as the reduction of acetophenones, and these complexes are clearly worth investigating, where applicable.

Non-chiral reduction of ketones can be accomplished by sodium borohydride or lithium aluminium hydride (LAH), and various chiral auxiliaries have been devised that make such reductions stereoselective [29, 30]. Such methods are not general, but can give high e.e. values in some cases, such as the reduction of acetophenone derivatives and related compounds. Sodium borohydride is much safer to handle than LAH, since its reaction with water is less violent, so it is preferred for industrial use, but LAH can be purchased in THF-soluble bags to ease handling dangers. LAH is more expensive than sodium borohydride, and also produces large quantities of lithium and aluminium salts, so the environmental impact of the large-scale use of this reagent needs to be considered.

A widely applicable method for asymmetric ketone reduction is the use of borane (typically, as the disulfide complex) in combination with oxazaborolidine catalysts. 2-Methyl-CBS-oxazaborolidines (Fig. 1.23, **au**) and 2-methoxy-oxazaborolidines (**av**, the latter produced either 'in situ' or shortly before the reduction from chiral amino alcohols and trimethylborate) are the main catalysts used [31, 32, 33], both isomers being available. The oxazaborolidines give enantioselective reduction of a wide range of ketones. This method has the advantage of not involving heavy metal catalysts. The use of the borane dimethyl sulfide complex creates an unpleasant odour, as the dimethyl sulfide is freed during the reaction, and a bleach scrubber may be necessary on a large scale to absorb this highly volatile liquid. The typical precursor to the oxazaborolidines is R or S α,α-diphenylprolinol (**aw**, also known as α,α-diphenyl-2-pyrrolidinemethanol). Unfortunately, this compound, being somewhat psychoactive, is occasionally abused, despite its serious side-effects [34]. The misuse of the compound has led to increased regulation and subsequent intermittent supply problems in some countries.

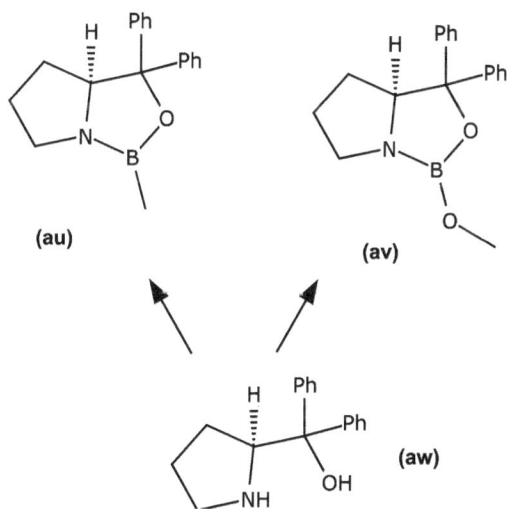

Fig. 1.23: Oxazaborolidine reduction catalysts and α,α-diphenylprolinol (**aw**).

A number of biotransformation reactions have been used for stereoselective ketone reduction. Isolated enzymes can be employed, although relatively expensive cofactors, such as NADPH, are needed. Cofactor recycling can be accomplished by the use of another enzyme, such as a glucosedehydrogenase (GDH). The latter enzyme has been used [35] in combination with a ketoreductase (KRED) to convert a β-ketoester (Fig. 1.24 **ax)** to a chiral alcohol **(ay)** with 99.9% e.e.; the KRED reduces the keto group while glucose is simultaneously oxidised by the GDH to enable efficient NADPH recycling by regenerating this cofactor.

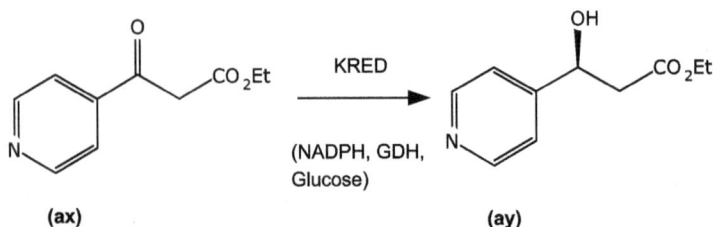

Fig. 1.24: Ketone reduction with a KRED enzyme.

Recombinant enzymes can also be used for reductions, such as modified ketoreductases (KREDS)[36], and these have been tailored to reduce ketones where native enzymes fail, greatly extending the scope of biotransformation methods.

Whole cell methods are typically more robust than isolated enzymes and can give good yields without the need for added cofactors. The reduction [37] of simple β-ketoesters to give (S)-β-hydroxyesters using baker's yeast has been known for over a century, but is far from general. The reaction has been extended to various ketones containing an aryllsulfone, arylsulfoxide or arylsulfide group at the α-position [38]. Yeast cells are often encapsulated in a suitable matrix, such as sodium alginate, in order to give easier separation from the product. The stereoselectivity of yeast can be increased or even reversed by genetic modifications [39]. Other genetically modified organisms, such as *E. coli*, can be used to reduce various ketones to chiral alcohols [40, 41].

The above survey of stereoselective ketone reduction is not comprehensive, but shows the main options that would need to be considered for accomplishing this transformation. If more than one of these methods shows initial promise in the laboratory, a detailed comparison involving cost, yield, product purity, environmental considerations, etc. would need to be carried out. Such comparisons between methods apply to most transformations; there is seldom only a single reagent that does the job, and one that gives too many side-products or is awkward to use on a large scale can delay process development. Such delays may eat into the patent life of a new compound, which can severely affect a company's profitability. Robustness and ease of development are important considerations, even trumping the theoretical possibility of increased yield on occasions. With an existing compound, where price is the determining factor, there may be more of an opportunity to spend time exploring more 'unusual' methods or reagents, if there is a promise of an eventual overall lowering of costs.

1.7 Solvent selection

Once a reagent has been chosen, a suitable solvent is often needed, although some reactions involving liquid reactants and products (or low-melting stable solids that can be used above their melting points) may not need a solvent at all. When a solvent is required, industrial reactions are typically run at relatively high concentrations, in order to minimise solvent costs and distillation time, while maximising the output from a given reactor. Starting material concentrations in the range of 10% w/v to 30% w/v are typical.

Some laboratory solvents are rarely used on a large scale. For example, diethyl ether, with a boiling point of around 35 °C [42], is likely to end up as a gas on hot summer days. It can easily be ignited by hot objects or surfaces, even without a flame being present. To make matters worse, it also has a potential to form explosive peroxides. Higher boiling ethers such as THF, 2-methyl-THF and diisopropyl ether (DIPE) are preferred, but peroxides may still be a problem, and testing is required to avoid the dangers of an explosion (particularly likely when removing an ether by distillation, since the peroxides become concentrated in the distillation pot). Stabilisers are often added to commercial ethers to inhibit peroxide formation, but can contaminate products, particularly when large quantities of solvent are used for chromatographic separations.

In addition to low-boiling solvents, water-miscible solvents tend to be avoided if non-miscible alternatives can do the job. Work-up is easier if washing with water is feasible, and in many cases the product does not need to be isolated, but can be carried on in solution to the next reaction stage, after suitable washes to remove inorganic compounds and unwanted impurities. Common immiscible solvents include toluene, xylenes, isopropyl acetate (i-PrOAc), n-butanol and 2-methyl-THF. All of these form azeotropes with water, so will remove traces of water when distilled off during work-up, thus avoiding the need for drying agents. Such solvents are used in a vast array of industrial processes and can be recycled by distillation, if necessary, decreasing the overall cost and the environmental impact of the process.

Polar aprotic solvents such as DMF, DMSO and NMP are required for some transformations. Such solvents are water miscible, so the addition of a second immiscible solvent is required if washing with aqueous phases is needed at the end of the reaction. Distillation of DMF and DMSO from reaction mixtures raises stability issues [43] and requires a high vacuum, so the recycling of these solvents is less common.

Chlorinated solvents are frowned upon due to their undesirable environmental properties, although there are a number of reactions, such as chlorinations of heterocycles with PCl_5 or Friedel-Crafts reactions, where they are hard to avoid. Many laboratory solvent-free methods are not suitable for use on a large-scale due to poor heat removal. In some instances, the starting material can be used as a solvent, such as Friedel-Crafts reactions with toluene or xylenes, but this is only practical with a few substrates. The use of methane sulfonic anhydride enables the acylation with carboxylic acids (via the mixed anhydride, which is formed in-situ) of some simple aromatics, such as toluene or mesitylene, with neither a Lewis acid catalyst nor solvent being required, other than the aromatic reactant [44]. However, this method is not general, and there is no universal method for the avoidance of chlorinated solvents in Friedel-Crafts reactions. Nitro compounds, such as nitromethane or nitrobenzene, have been used in some cases although they are not recommended, since nitromethane is explosive [45] and nitrobenzene highly toxic [46].

Water is considered the ideal solvent, although most starting materials are not soluble enough to avoid the formation of a two-phase system, which often leads to slow rates of reaction. Phase-transfer reagents can speed up reactions in water, but may complicate work-up. Even when reactions can be carried out in water, the work-up is often far from straightforward and may require the addition of a solvent to enable the recovery of the product without loss of yield, thus negating the environmental advantages of using water. Salting-out procedures may be helpful for product isolation from water.

Ionic liquids can be used to carry out a variety of reactions, sometimes giving superior yields to conventional organic solvents [47, 48]. Ionic liquids are typically expensive, so practical recycling procedures are required to make their use economically viable. Being non-volatile, there are no losses of ionic liquids by evaporation or any danger from harmful or flammable vapours. Of course, significant losses may occur when removing the product or impurities by extraction. Ionic liquids generally have a low solubility in alkanes, so an alkane extraction can be used to remove non-polar impurities or products. However, their solubility in water can lead to unwanted contamination of aqueous washes, giving rise to environmental problems of disposal.

1.8 Costing

Potential synthetic routes need to be costed and compared. Bulk prices for starting materials and reagents are nearly always lower than laboratory catalogue prices, often by a factor of 10 or so [49]. Quotes from manufacturers are needed for more accurate estimation of costs. A 'feel' for likely yields is useful for costing estimates: a simple esterification could be estimated to be likely to give a 90% yield, while for an oxidation with a variety of likely impurities, 70% might be a better guess. Once some laboratory work has been carried out, more accurate yields should be available. Capital costs have to be considered, and they greatly depend on which vessels are required. A process that can more or less fit into existing plant will clearly have lower capital costs than one in which a complicated special plant needs to be constructed. Other costs such as labour, overheads and energy can vary widely between countries, and these differences may be crucial in decisions regarding the siting of plants. Close collaboration with chemical engineers is vital in order to

obtain reasonably accurate costings. Spreadsheets are normally used for cost estimates, since they can be easily updated as prices and yields change. Table 1.1
shows a typical rough initial costing calculation for a single stage using hypothetical compounds and prices. The calculation assumes 90% of the solvent is recycled
and reused in the next batch. Without solvent recycling, the cost contribution of the
solvent increases from £0.71 per kg of product to £7.13 per kg of product, greatly
adding to the overall cost.

Tab. 1.1: Initial rough costing for a single (hypothetical) stage.

		Molar ratio	Mol. wt.	kg used per kg prod.	Price £/Kg	Cost per kg prod.
Reactant A		1.00	178.23	0.701	£5.10	£3.58
Reactant B		1.00	122.16	0.481	£4.20	£2.02
Catalyst		0.05	190.22	0.037	£1.80	£0.07
Product		0.90	282.37			
Molar yield	90%					
Solvent	Solvent	%	Unrecycled	Unrecycled		
	L/kg React. A	Recycled	L/kg A	L/kg product	Price £/L	
	10.00	90%	1.00	1.426	£0.50	£0.71
					Capital	£2.00
					Other costs	£2.10
				Cost/kg prod.	Total	**£10.48**

Once a suitable route has been chosen and shown to work in the lab, your thoughts
can turn to scale up. The first consideration needs to be the potential hazards in the
scale-up of the resulting process; it's always wise to consider these upfront rather
than have them forcefully shown to you, either by your colleagues or by the reactions themselves, some months down the line. These hazards will be discussed in
the following chapter.

Chapter 2
Scale-up hazards

2.1 Overview of hazards

As reactions are carried out on a larger and larger scale, the potential dangers also increase. Identifying hazards and reducing the associated risk is a vital part of process development. Neglect of plant safety has sadly led to many preventable injuries and deaths, so this is a topic that all process chemists should understand in depth. However, on the whole, the safety of chemical processes has improved over the years. For example, in the 1950s, ICI (at that time the leading chemical company in the UK) ran fluorination processes [50] that became extremely unstable if the temperature was too low (a not uncommon occurrence). Such processes would be avoided at the design stage today, where processes are devised to 'fail-safe', so that any removal of heat, stirring, cooling, etc. will not lead to a serious incident. Nowadays, most process development chemists go through their careers without experiencing a large-scale fire or explosion. Large companies often have a dedicated process safety laboratory (PSL), which has overall responsibility for process safety, but in many establishments it is part of the job of the process development chemists designing the process, in collaboration with chemical engineers, analytical chemists and plant managers, to carry out the necessary studies in order to ensure process safety.

Broadly, process hazards can be divided into explosions, over-pressurisation, fires, toxicity and corrosive hazards, all of which have to be assessed during the risk assessment of a process. Some hazards are not immediate, such as the slow corrosion of a vessel or pipework, which can lead to catastrophic failure after several batches, and may not be apparent from laboratory runs in standard glassware. In some cases, hazards can be avoided by the substitution of a less toxic or unstable reagent. Where it is not practicable to avoid hazardous reagents or intermediates, the amount present at any one time can sometimes be reduced by the use of continuous reactors. Whatever the plant set-up, suitable personal protective equipment (PPE) is essential for minimising risks. Apart from hazards associated with processes, other hazards, such as those due to manual handling, electricity and tripping, exist on plants and should not be neglected. Back injuries due to lifting can easily arise in plants, particularly when inexperienced staff try to move heavy objects without proper assessment of the risks.

https://doi.org/10.1515/9783110717877-003

2.2 Exothermic reactions and calorimetry

Exothermic reactions are one of the greatest hazards that need to be considered on scale-up. The heat produced from a reaction increases in proportion to the volume, while the cooling from a jacket increases in proportion to the surface area. Since the volume of a spherical reactor is proportional to the radius cubed, while the surface area is proportional to the square of the radius, the ability of a cooling jacket to remove heat decreases as the size of a reactor increases. Consider a laboratory spherical flask of radius 0.1 m – the volume will be $(4/3)\pi r^3 = 0.0041888$ m^3, the surface area $4\pi r^2 = 0.12566$ m^2 and the ratio of volume to surface area will be ca. 0.033. For a pilot plant reactor of radius 1 m, the volume will be 4.1888 m^3, the surface area 12.566 m^2 and the ratio of volume to surface area will be ca. 0.33, ten times greater. This is a simplified model, but it shows that increasing the scale makes a cooling jacket less and less effective. Overheating is particularly dangerous when unstable substances are present, or when the reaction mixture reaches the boiling point of the solvent, causing sudden over-pressurisation. Reactors are typically fitted with a 'bursting disc', which is designed to break under pressure, avoiding catastrophic destruction and instead ignominiously depositing the reaction mixture into a suitable overflow tank (also known as a catch tank or dump tank).

Initial examination of thermal hazards often involves running differential scanning calorimetry (DSC) on starting materials, reagents, products and isolable intermediates [51]. In DSC, a weighed sample of a few milligrams is sealed in a small metal 'pan'. An empty reference pan is used for comparison. The two pans are gradually heated in the instrument, and a graph is produced showing the differences in heat flow between the two pans against temperature. The scan will show the temperature of any melting (endothermic peak if no decomposition), other phase transition or exotherm. Figure 2.1 shows a typical (hypothetical) endothermic melting point peak trace.

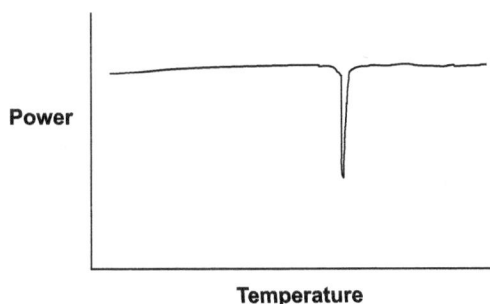

Fig. 2.1: DSC (hypothetical) of a melting point.

DSC examination of solids should be carried out both in the dry form and when wet with the solvents used. Such studies are particularly important before any large-scale drying is carried out. Phase transitions between different polymorphic forms are not uncommon on heating solids and may be both exothermic and occur significantly below the melting point. Any significant exotherms at temperatures relatively close to the temperature of operation are a cause for concern and should trigger a much more detailed investigation, or even a change of route. It should be noted that the onset of an exotherm in a DSC typically occurs at a somewhat higher temperature than in the bulk solid, due to the very small size of DSC samples. Figure 2.2 shows a typical (hypothetical) exotherm trace.

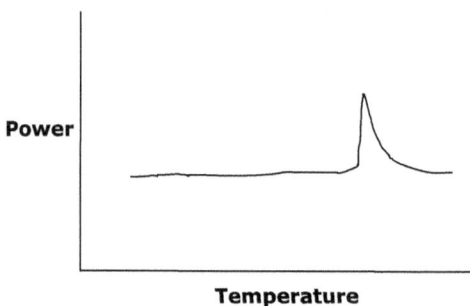

Fig. 2.2: DSC (hypothetical) of an exotherm.

Accurate measurement of the magnitude of significant reaction exotherms is vital and needs to be carried out on the actual reaction mixtures. Simply mixing the reactants and looking at the temperature rise will lead to underestimation of exotherms, due to heat losses to the environment. However, there are a number of commercial laboratory reaction calorimeters available, such as Syrris's Atlas HD Reaction Calorimeter, Mettler Toledo's RC1mx or H.E.L.'s Simular I Heat-Flow Reaction Calorimeter. These will give accurate thermodynamic data, including heats of reaction, if used according to the manufacturer's instructions. Such calorimeters can be used for heat flow calorimetry (HFC), in which the heat flow through the reactor wall is measured. Calibration runs are required, so the process can be lengthy. An alternative is power compensation calorimetry (PCC), which is supported by some reaction calorimeters. In this mode, a heater is used to maintain isothermal conditions by compensating for the cooling effect of the jacket. This avoids the need for calibration runs, but there can be problems from hot spots around the heater. Some consultants and academic laboratories offer calorimetric services if a company does not possess the necessary equipment.

If necessary, rough estimates of heat losses to the environment can be made by carrying out a reaction where the heat of reaction is known (such as adding an acid to an alkali) on the same scale and in the same equipment in which the reaction of interest is tested. Such methods do not give very accurate results, but are better than simply guessing.

As well as the reaction exotherms that arise during normal processing, the adiabatic temperature rise [52] is also of interest. This is the temperature produced with no heat loss to the environment and all the initial starting materials and reagents being present from the beginning (no controlled addition of reagents). Such a temperature rise gives a worst-case scenario, with all the inputs added to the vessel and the cooling having failed, and is useful to flag up reactions that need detailed evaluation. Small adiabatic temperature rises of less than 50 °C are generally considered of low severity (except when the rise would take the reaction mixture above the boiling point of the solvent), while an increasing degree of concern is generated as the rise moves above this point [53]. The adiabatic temperature rise can be estimated experimentally or calculated. Where potential runaway reactions are flagged up by DSC or reaction calorimetry, accelerating rate calorimetry (ARC), which gives more detailed quantitative information on onset temperatures, heat generation and potential pressure changes under adiabatic conditions, can be advantageously carried out.

Intrinsically unstable compounds have a self-accelerating decomposition temperature (SADT), which is the temperature at which the bulk solid starts heating of its own accord. The SADT needs to be determined for safe operation. For example, using a variety of methods, including the direct heating of large samples and theoretical calculations based on DSC data, the SADT of the azo initiator AZDN (AIBN, 2,2′-azobisisobutyronitrile) (Fig. 2.3) was determined as being between 47 and 55 °C for the bulk solid [54]. This is not one for heating in the oven.

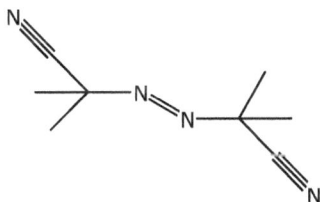

Fig. 2.3: AZDN (AIBN), an unstable azo initiator.

For batch reactions, exotherms are typically controlled by the slow, controlled addition of a reagent or starting material, sometimes referred to as a 'semi-batch' reaction. Clearly, this addition is more easily carried out if the substance to be added is a liquid, which can be pumped in or added from a graduated glass header vessel. Solids are conveniently added in solution, although in some cases this is not possible. Various screw feeders have been devised for the slow addition of solids, but none are immune from blockages. In practice, solids may have to be added via the vessel manway, with the operator suitably attired in an air suit. If this procedure is adopted, great care is needed to ensure that minimum foaming occurs; otherwise, the vessel contents may escape via the manway. I know of one instance, many decades ago, where a somewhat bulky chemist prevented such an escape by putting their full weight on the manway cover, while their colleague quickly bolted it into place. Such procedures are not recommended.

The slow addition needed to control exotherms adds considerably to reaction time, so an addition that is carried out in minutes in the laboratory becomes hours in the pilot plant. Obviously, a prolonged addition time may affect the chemistry, so it is vital that such times are estimated prior to pilot plant runs and laboratory experiments carried out to ensure that the chemistry is not adversely affected. Again, collaboration with chemical engineers is vital. Looking at how long other exothermic reactions have taken in a particular pilot plant vessel can also help with achieving reasonable estimates for addition time.

It might be thought this the highest possible cooling should be used on exothermic reactions, but this is not always the case. There is a particular danger of a reaction becoming so slow that dangerous amounts of reactants build up. The eventual reaction can now be far harder to control than a steady reaction at a higher temperature. Reactions that do not necessarily start immediately, but instead have an induction period, are particularly hazardous. The formation of Grignard reagents on a large scale requires particular care. Typically, the organic halide is slowly added to a slurry of magnesium turnings in a suitable solvent, often THF or 2-MeTHF. All sorts of catalysts, such as traces of iodine, dibromoethane or iodomethane, have been used to ensure the smooth formation of Grignard reagents, but the most reliable method is to add a catalytic amount of the required Grignard reagent that has been previously formed in the laboratory. Often, this is added after a small fraction of the halide has been added to the magnesium. Careful attention to reactor temperature traces should show an exotherm when a Grignard reagent begins to form. The bulk of the organic halide should not be added until such an exotherm is noted.

Occasionally, emergency actions are required to control unexpected exotherms on a pilot plant, provided the rise in temperature is not rapid enough to warrant immediate evacuation. Such exotherms may occur if there is a failure of the cooling system or a change to the usual course of a reaction (such as an oil unexpectedly turning solid and making the mixture too viscous). It is sometimes possible to reduce the temperature by adding additional solvent. If the solvent is at a lower temperature than the reaction mixture it will tend to have a cooling effect, while in most cases the increased dilution will also tend to slow the rate of reaction. This procedure is only useful if the reactor is not already full, so it is a good argument for allowing plenty of space above a reaction mixture, which also allows room for some foaming to occur. It is not good practice to run pilot plant batch reactions with completely full reactors. Water can also be pumped into a reactor if none of the substances present react or give exotherms. An emergency quench may also be employed, in which the reaction mixture is dropped into a larger reactor containing a large excess of a suitable quench medium, such as water or an alcohol, as appropriate.

Effective mixing is essential for the control of exothermic reactions, and it's important to use the correct agitator. For a batch reaction involving solids that tend to sink to the bottom of the reactor, a 'U-shaped' anchor agitator is typically preferred, with minimum 'dead-space' in which solid can accumulate. For a reaction involving the rapid mixing of liquids, a fast propeller-type agitator would normally be better.

Some exothermic reactions are better controlled by using a continuous reactor, particularly where reactions are fast and do not involve solids, which may lead to blockages. Due to the smaller size of continuous reactors, much greater cooling is possible. Some reactions can be carried out in metal tubes, where the liquid reagents mix as they are pumped through and external cooling is employed. For other continuous reactions, an actual vessel is needed, but sufficient mixing can be obtained by the fast injection of gas (either as a reactant, such as chlorine, or an inert gas, such as nitrogen) to make an agitator redundant. Without the need to accommodate an agitator, internal metal cooling pipes can be used, greatly increasing the available cooling. Such a reactor is shown in Fig. 2.4, with chlorine and liquid reactant entering through the bottom of the reactor, and the liquid chlorinated product overflowing into an outlet tube at the top of reactor, while hydrogen chloride gas is vented from the top. The diagram shows only two cooling tubes for clarity, although more would be used in practice. These would be made of tantalum, which is relatively resistant to a mixture of chlorine and hydrogen chloride. Horizontal baffles might also be used to ensure efficient mixing.

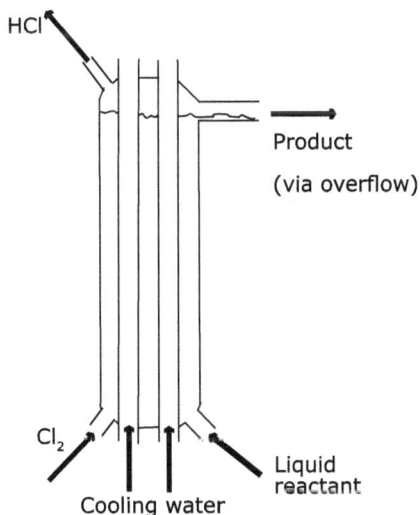

Fig. 2.4: Chlorination in a vertical continuous reactor.

2.3 Unstable reagents, catalysts, intermediates and by-products

Reagent and catalyst stability are just as important as those of reactants and products. Hydrogenations and other metal-catalysed reactions are particularly notorious for causing fires, the dry catalysts being liable to cause ignition of hydrogen or solvent vapours. Filtration of the catalyst after the completion of the reaction is often the most dangerous stage, since the catalyst may dry out on the filter if care is not taken. Keeping the catalyst wet with water and under a nitrogen atmosphere during and after its filtration helps prevent ignition. I find running a video of the Hindenburg disaster on a control room computer helps focus minds on safety during hydrogenations.

Sodium and lithium metals are typically used in Birch reductions, but both react with water (very vigorously in the case of sodium) and tarnish in air. For some Birch reductions, less reactive calcium metal can be used instead [55]. Liquid ammonia is used as solvent in Birch reductions, and this toxic substance requires careful handling. Recent laboratory work from the Baran group [56] has demonstrated an electrochemical alternative that does not require liquid ammonia and reduces a wide range of substrates, such as phenylethanol (Fig. 2.5), although it remains to be seen if this method will be commercialised.

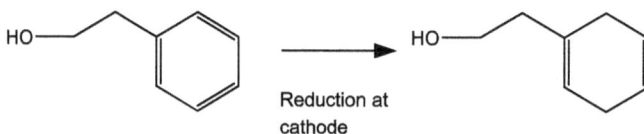

Fig. 2.5: Electrochemical reduction of phenylethanol.

Diborane is a highly flammable, toxic gaseous reducing reagent, which is normally used as its more stable dimethyl sulphide complex, although the THF adduct is also sometimes employed. Both complexes are commercially available in solution, giving reagents that are much safer to use than the neat gas. The auto-ignition temperature (AIT) for pure diborane in air was recently measured as 136–139 °C [57].

There are cases in which both the reactant and product are reasonably stable, but the reaction proceeds via an unstable intermediate. One common case is the reduction of aromatic nitro compounds to anilines, which goes via an unstable hydroxylamine compound [58]. Careful optimisation of the reaction conditions is needed to prevent accumulation of the hydroxylamine derivative (Fig. 2.6).

Ozonolysis reactions on double bonds lead to the formation of unstable ozonides, which are then decomposed, usually by being reduced to carbonyl compounds or alcohols (Fig. 2.7). However, incomplete decomposition of the ozonides can lead to explosions during work-up. Ozone itself, which is normally produced in oxygen by an ozone generator, is an extremely hazardous gas, unsuitable for bulk

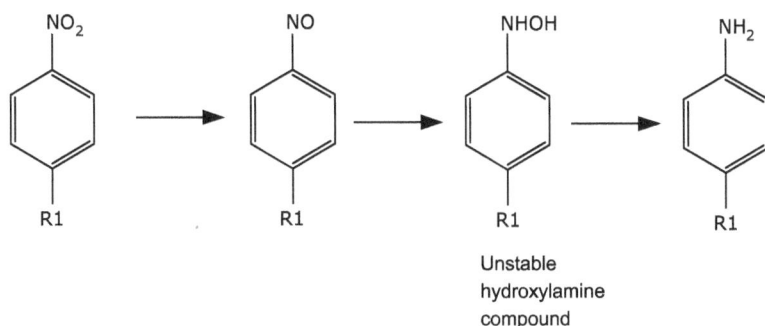

Fig. 2.6: Nitro reduction via unstable hydroxylamine.

batch reactions. Continuous flow reactors have been used to reduce the dangers of ozonolysis by companies such as Lonza [59].

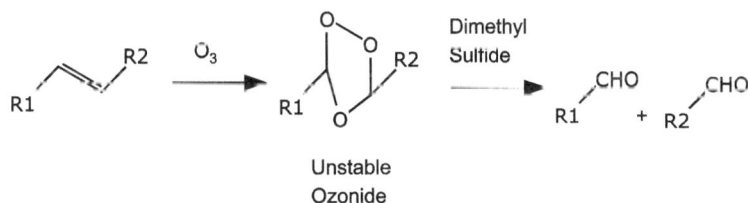

Fig. 2.7: Ozonolysis via unstable ozonide.

Diazonium salts are another group of unstable intermediates, which are usually formed in situ, below 5 °C, prior to undergoing further reactions to give a wide variety of products. Tight control of the operating conditions is essential [60]. A number of explosions have been reported over the years, both in laboratories and on plants [61–63]. Continuous flow methods can minimise the amount of diazonium salt present at any one time [64–66]. The latter reference involves two unstable intermediates, the diazonium salt (Fig. 2.8 **a**) and the azide (**b**), both kept at low levels by the use of flow chemistry.

Unstable compounds may occur as unwanted by-products and must be safely removed. An example is during aromatic nitration, where typically mononitration is desired, but traces of unstable di- and tri-nitro compounds may form. Subsequent distillation of the mononitro compound can lead to dangerous concentrations of the di- and tri-nitro species. Such a procedure was the cause of the 1992 Hickson and Welch disaster, in which five people lost their lives. Tarry residues from a nitration distillation still had been allowed to accumulate over many years. An attempt was made to remove the residues using steam softening and a metal rake, which eventually led to a jet fire, as two immense jets of burning vapours projected out of the vessel [67]. Such incidents can be avoided by proper safety

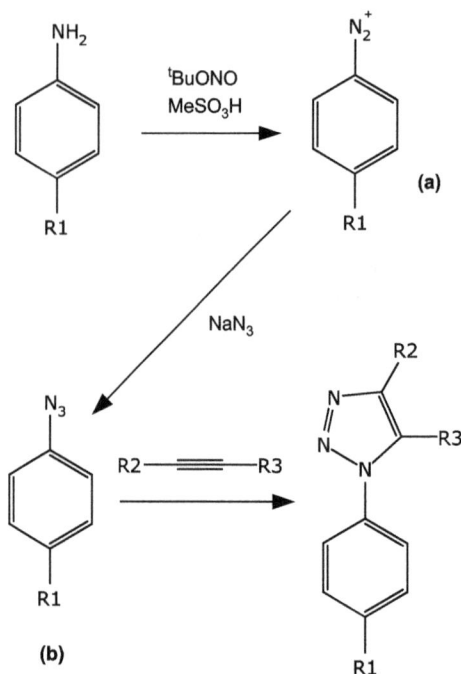

Fig. 2.8: Continuous formation of diazonium salt (**a**) and azide (**b**).

evaluations before any work is carried out and the regular removal of distillation residues, so that they are not allowed to accumulate.

2.4 Solvent hazards

The stability of solvents needs to be carefully considered. There are no totally inert solvents: all will undergo reactions under some conditions. Even when a solvent is stable under the reaction conditions, decomposition can occur during work-up or recovery for recycling.

The decomposition of DMSO [68, 69] has been the cause of a number of laboratory and plant incidents over the years. Acid-catalysed autocatalytic decomposition can occur at temperatures below its boiling point even under an inert atmosphere. The decomposition can give rise to formic, acetic, methanesulphonic and sulphuric acids, which serve to increase the rate of decomposition. Strong bases remove a proton from DMSO to form the dimsyl anion, which has a number of synthetic uses. However, the reaction of sodium hydride and DMSO can cause a run-away exotherm and the explosive formation of gases [70]. Similar hazards can occur with sodium hydride and DMF or DMAc. Although most process chemists are aware of the danger of such mixtures, a few large-scale reactions are still reported in the literature.

Acetonitrile is sometimes assumed to be an inert solvent, possibly because of its widespread use in HPLC. However, it can react with nucleophiles under conditions of both high and low pH. Acetonitrile and aqueous sodium hydroxide react exothermically and have the potential for a runaway reaction [71]. Under acidic conditions, alcohols can add to acetonitrile, giving an alkyl imidate salt via the Pinner reaction (Fig. 2.9). Subsequent hydrolysis can lead to the rapid evolution of ammonia.

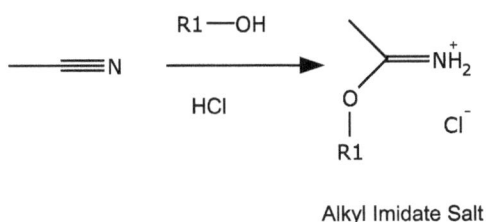

Alkyl Imidate Salt

Fig. 2.9: Pinner reaction of acetonitrile.

Ether solvents, such as THF, dioxane and diisopropyl ether readily form unstable peroxides, as do a few hydrocarbons, such as cumene. Peroxides are particularly hazardous when the solvent is removed by distillation, where the peroxides become concentrated and can explode. Commercial THF and diisopropyl ether are typically sold with butylated hydroxyl toluene (BHT) as an inhibitor. However, BHT can potentially contaminate products, particularly when they are isolated as an oil or when large quantities of solvent are removed following purification by preparative column chromatography. Inhibitor-free tetrahydrofuran is available, but should be tested for peroxides and always used fresh. This is not something you want to leave at the back of the warehouse for your next pilot plant campaign.

Small amounts of solids in solvents can lead to the clogging of filters and other equipment, which can be a particular problem with continuous processes. For example, small amounts of poly(THF) impurity from a particular batch of THF have blocked flow equipment on a plant-scale process [72]. Solids, including product, can sometimes drop out of solution on standing, particularly if drums containing solids in organic solution are left outside on cold winter nights.

In most plants, the flammability of solvents is controlled by the nitrogen inerting of vessels. The use of inert gases undoubtedly makes processes safer, but can lead to dangers, including asphyxiation when workers enter confined spaces containing an inert gas, which can happen during vessel maintenance or due to leaks in small rooms. Permit to work systems should dictate that the nitrogen supply to a vessel is blanked off and the vessel purged with air, prior to entry into the vessel. In addition to inerting procedures, solvent fires in plants are also prevented by reducing the likelihood of sparks. Antistatic measures, such as earth (grounding) connections on drums and pumps during solvent transfer and wearing antistatic boots, are essential safety precautions. Plant electronic equipment needs to be 'Ex-rated' (i.e. designed to appropriate specifications, so as not to be a source of sparks). As well as solvents, dusty organic solids can also pose an explosion risk and require similar precautions.

2.5 Incompatible substances and unintended gas formation

It is important that the substances and solvents used in a reaction and work-up are mutually compatible. Some incompatibilities are obvious, such as acids and bases. Less obvious ones include acetone and chloroform, which interact with a large exotherm in the presence of base, where dichlorocarbene can be formed and goes on to react with the acetone. Even in the absence of base, there is some exotherm due to hydrogen bonding between the hydrogen in the chloroform and the oxygen of the acetone.

Hypothetical scenario 1

After running a chromatography column, a chemist ends up with 100 L of mixed dichloromethane and acetone. They add this to an empty waste drum and label it as 'waste chlorinated solvent'. The drum sits out in the yard for a while. A second chemist adds a further 80 L of chloroform, followed by a few litres of chlorobenzene contaminated with aqueous sodium hydroxide. They tighten up the drum cap and walk away, narrowly avoiding shrapnel from the exploding drum.

Such dangers show the need to actually list the contents of waste drums, as general 'chlorinated solvent' or 'non-chlorinated' solvent labels are not appropriate when large quantities of waste are being disposed of.

'Bretherick's Handbook of Reactive Chemical Hazards' [73] is an invaluable source about chemical incompatibilities, and should always be consulted by process development chemists. It is available both in book form (currently on its 8th edition) and online. Another useful volume is Stoessel's 'Thermal Safety of Chemical Processes' [74], which covers the thermal hazards associated with different types of reactors, while also including numerous case studies.

Some reagents, such as hydrogen peroxide, give rise to oxygen gas. This can be diluted to a safe level with a strong nitrogen flow and slow reagent addition. Oxygen meters on the reactor and gas outflow line ensure that levels are kept low.

Although the reaction of carbonate salts and acids seems straightforward, in practice, it can be delayed due to slow reactions of weak acids [75] or suspensions of undissolved carbonate.

Hypothetical scenario 2

A reaction is carried out for the first time on a pilot plant. Two process development chemists are present, along with their team leader, two pilot plant staff, a chemical engineer and the pilot plant manager. The head of process development puts in an appearance to check that everything is going as expected. Several samples are taken, and all temperature changes are carefully noted and recorded. Great care is taken with every step. At the end of the shift, the product is safely in the oven, and the initial indications of purity and yield are excellent. Everybody goes home congratulating themselves on a job well done. Two inexperienced members of staff arrive for the night shift. All they have to do is to drum off aqueous layers and ready the plant for batch two. The aqueous layers contain carbonate washes and are quickly put into drums, which are left outside in the yard with their caps tightly shut while other tasks are attended to. The carbonate washes contain a weak acid that slowly leads to the formation of carbon dioxide. The drums finally explode just as the morning shift arrives.

It is therefore important to leave the caps on carbonate and bicarbonate waste drums loose to prevent pressurisation on standing. Labels should be attached warning of loose caps, so the drums are not moved in this state. More generally, this scenario shows the importance of considering hazards during routine steps, such as waste storage and cleaning, rather than just those steps involving the reaction and work-up. For example, the cleaning of flexible hosing can give rise to incidents, particularly when cleaning is left to a subsequent shift, when the staff might be unaware that corrosive or toxic substances may be left in the hoses. Clear labelling of equipment and detailed, written shift changeover notes are required to prevent such incidents.

2.6 Toxic substances – their control and substitution

Apart from fires and explosions, the toxicity of substances needs to be considered. Sometimes toxic substances can be replaced by less dangerous alternatives. In addition, the amount of a toxic substance in use at any one time can often be reduced by using continuous as opposed to batch reactors, although these are not always practical. Some toxic substances can be generated in situ. For example, diphosgene (trichloromethyl chloroformate) or triphosgene (bis(trichloromethyl)carbonate, BTC) (Fig. 2.10) can be used instead of the extremely toxic gas phosgene [76]. However, both these alternative reagents are in equilibrium with phosgene gas, so they need to be treated with great caution. In practice, diphosgene, being a liquid, is easier to handle on a plant than triphosgene, which is a solid. Diphosgene can readily be produced by the photochemical chlorination of methyl formate or methyl chloroformate [77]. Diphosgene can be used in the preparation of a wide range of isocyanates [78], which may further react to give carbonates, carbamates and related compounds. Since phosgene was used to terrible effect as a chemical weapon in the First World War, its use is tightly controlled, although the exact licensing regulations applicable vary from country to country. A milder phosgene analogue is 1,1-carbonyldiimidazole (CDI), which is more expensive and less reactive than phosgene, but can be used to good effect in making various compounds such as organic carbonates [79]. In some syntheses, dimethyl carbonate (DMC) can be used as a benign phosgene replacement [80].

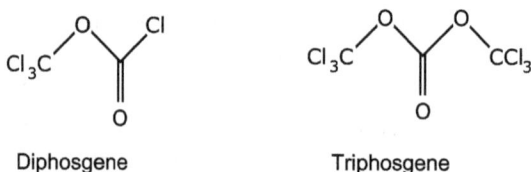

Diphosgene   Triphosgene

Fig. 2.10: Diphosgene and triphosgene.

N-Methyl carbamate insecticides can be made by reacting phenolic compounds with phosgene or its precursors to give a chloroformate (Fig. 2.11 **c**), and then reacting the latter compound with methylamine to give the N-methyl carbamate (**d**). The direct reaction of phenols and methyl isocyanate is an alternative, but since the Bhopal disaster in 1984, caused by the large-scale emission of the latter compound, many producers have avoided its use. Since methyl isocyanate is normally produced from phosgene and methylamine, the use of phosgene is not necessarily avoided by employing methyl isocyanate. The production of carbamates from carbon dioxide has been achieved in the laboratory [81] under relatively high temperatures and pressures, although at the time of writing such processes are rarely used industrially for insecticide production.

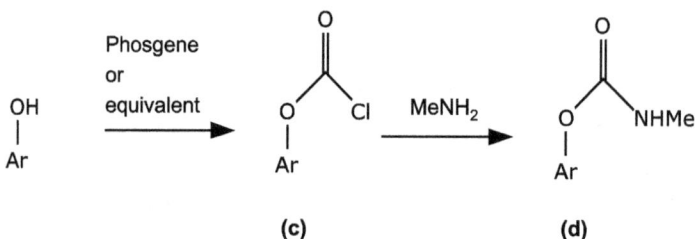

(c)   **(d)**

Fig. 2.11: Carbamate insecticides (**d**) from chloroformate intermediate (**c**) and methylamine.

Chlorine gas is sometimes replaced by the use of hydrochloric acid and an oxidising agent, such as sodium chlorate [82] or hydrogen peroxide, with a suitable catalyst [83]. Such procedures avoid the hazards of using chlorine gas, although care must be taken to control oxygen levels in the reactor, usually by slow addition of the oxidising agent. Similarly, bromine can be replaced by the use of HBr and an oxidising agent, such as hydrogen peroxide or sodium bromate. An example of the latter is a continuous photochemical benzylic bromination [84] run with an output of up to 4.1 kg/h.

Fluorine itself is seldom used for fine chemical synthesis, due to its extreme reactivity and toxicity. Hydrogen fluoride is sometimes employed, but is notorious for its corrosivity and toxicity [85, 86]; it is one of the few substances that corrode glass. Reactions involving fluorides can form traces of hydrogen fluoride during reactions or work-up if the pH becomes too low, both etching glass-lined vessels and presenting a serious toxicity hazard. Appropriate PPE must be employed if there is any risk of hydrogen fluoride being generated, while calcium gluconate gel should be available to treat any burns, prior to emergency removal to hospital.

Hydrogen cyanide is another dangerously toxic substance [87]. Toxic cyanide salts are sometimes employed for nitrile formation, and there is a risk of generating the even more hazardous hydrogen cyanide if the pH becomes too low during the reaction or work-up. At the time of writing, treatment with oxygen administered by trained first-aiders is recommended in the UK in cases of cyanide poisoning [88], prior to emergency removal to hospital, although guidelines vary from country to country.

Dimethyl sulphate is widely used as an industrial methylating agent, despite being carcinogenic, since it is inexpensive and gives high yields over a wide range of reactions. Much effort has been applied to finding suitable alternatives, although none is universal. DMC, although less reactive, is a useful alternative for various methylations [89, 90, 91]. Methanol itself can be used in some cases, such as the methylation of phenols, but very high temperatures of above 300 °C are required, which many compounds would not survive [92]. Of course, methylation and other alkylation reactions of amines are often carried out by reductive amination, which is a very common industrial reaction [93].

Osmium tetroxide or related osmium compounds are used as catalysts, in combination with an oxidant such as *N*-methylmorpholine *N*-oxide, for the *syn*-dihydroxylation of alkenes. However, osmium tetroxide is both volatile and highly toxic, as well as environmentally harmful, so it is far from ideal for plant use. The old reaction using permanganate tends to give over-oxidation, but can be modified by the use of phase-transfer catalysts to be more controllable, and other manganese compounds can also be used [94]. Another alternative is the use of cerium ammonium nitrate (CAN) and iodine [95]. Regardless of their lower toxicity compared to osmium tetroxide, permanganate and CAN are very strong oxidising agents, and care is needed when handling them on a large scale. Appropriate consideration is needed as to the handling of spills (absorption onto inert substances is needed, not oxidisable substances such as cotton wool or paper towels), quenching and disposal.

Toxic gases, such as chlorine, phosgene, hydrogen chloride and ammonia, are commonly used on plants, and some may be generated during reactions. Scrubbers are typically employed to prevent discharge to the atmosphere, with sodium hydroxide solution being used for acidic gases and a suitable acid, such as dilute sulphuric acid, for basic gases. It is important that scrubber capacity is adequate for the amount of gas emitted and that the pH is regularly checked. Neutral organic gases, such as methyl bromide (formed from some demethylation reactions) or vinyl bromide (a by-product from the use of 1,2-dibromoethane), may require scrubbers charged with a high boiling organic solvent, such as toluene or xylene. Aqueous sodium sulphide and a phase transfer catalyst have also been used to destroy methyl bromide [96]. Apart from its toxicity, methyl bromide is an ozone-depleting substance, so its removal is important for environmental reasons.

Lines or valves for toxic gases need to be checked for leaks as soon as the gas starts to flow. For chlorine or hydrogen chloride lines, the use of plastic 'squeeze bottles' containing cotton wool soaked with ammonia solution enables leaks to be quickly identified, as the white clouds of ammonium chloride are obvious. A variety of electronic gas detectors and gas detector tubes are available, such as those produced by Dräger, and should be employed whenever toxic gases are used or generated. Carbon monoxide is particularly hazardous, being both toxic and odourless, so great care should be employed when it is used or generated.

In general, workplace exposure limits to toxic substances are expressed as a time-weighted average. A long-term exposure limit refers to an 8-h exposure period, while a higher short-term exposure limit refers to exposure for 15 min. However, regulations vary from country to country, as does their enforcement. Exposure to toxic substances is controlled by suitable plant configuration. 'Elephant trunking' (local extraction via a wide flexible tube) is used to reduce exposure to harmful or toxic solvents, being placed over drum or vessel openings when solvent is added or removed. Enclosed filtration units, such as Nutsche filter-driers, can significantly reduce the potential for exposure during filtration and drying operations. PPE, such as air-fed suits or hoods, is required where exposure to toxic substances is possible. Decontamination, typically with a shower, is necessary before air suit removal. Sometimes other decontamination methods are needed; for example, aqueous ammonia solution can be used to destroy any carcinogenic dimethyl sulphate before PPE is removed. Dilute bleach can be used to oxidise thiol and sulphide compounds, reducing the unpleasant odour.

2.7 Corrosive substances

A wide range of highly corrosive substances are used in plants. Strong acids, such as concentrated sulphuric acid, are commonly used, and suitable PPE is essential when handling such substances. Corrosive alkalis, such as sodium hydroxide, require similar precautions. Air-fed suits may be appropriate, or a resistant apron, long gauntlets and visor. Many PPE manufacturers provide much useful information on compatibilities on their websites. It should be noted that standard nitrile gloves are not always the best choice, and other options, such as Viton® gloves, may be more appropriate for some substances.

In addition to protecting staff from injury, careful consideration has to be given to possible corrosion of vessels and pipework, which, if neglected, can lead to very costly replacement, if not failure and dangerous leaks. Standard glass-lined reactors are resistant to most chemicals, although corrosion by hydrogen fluoride, as mentioned previously, is a possible hazard. Exposure to hot concentrated caustic alkalis, such as sodium hydroxide solution, can also cause unacceptable glass corrosion over time. In the latter case, stainless steel is a possible alternative for a reactor, provided there is no other substance present that would attack it.

Stainless steel is subject to attack from a variety of acids, while it also corroded by chloride ions [97]. If a metal vessel or cooling tubes are required under such conditions (which may be necessary for rapid heat exchange, etc.), Hastelloy® alloy or tantalum may be employed. The latter expensive metal is useful for exothermic chlorination, since it will survive the onslaught of both chlorine and hydrogen chloride. However, even tantalum can be corroded under some conditions, such as bromination in the presence of pyridine and methanol [98].

2.8 HAZOP and minimising risk

Prior to a process being introduced onto a pilot plant, the effects of possible perturbations to the reaction conditions are formally examined. This procedure is called HAZOP (hazard and operability), and typically involves both chemists and chemical engineers sitting in front of a plant diagram for the process [99]. HAZOP is an essential element of the wider requirement for effective process safety management, which is the overall framework for managing process risk. HAZOP considers deviations systematically, so for the addition of a reagent to a vessel, the pertinent conditions might be: (1) Add too much, (2) Add too little, (3) Add too quickly, (4) Add too slowly, (5) Add at too high a temperature, (6) Add at too low a temperature, (7) Agitator failure during addition, (8) Nitrogen failure during addition or (9) Condenser leak during addition. Thus, lists of 'what if' questions are produced for every stage of the process. The answers to some of these might be readily available from previous laboratory work, but often further test reactions need to be run.

Most perturbations will not have catastrophic consequences, but sometimes changes to the process or plant configuration are needed so that risks are minimised. For example, the chlorination of a heterocyclic compound with phosphorus pentachloride under reflux might lead to a catastrophic reaction if water leaked from a glass condenser, giving a violent reaction and the generation of copious quantities of HCl gas. Assuming that reflux temperature is required, one possible solution would be to configure the plant so that the condenser receives vapour from the reaction, but is sited above an empty vessel, which would receive the water in event of a leak. Condensate could be pumped out of the empty vessel and recycled into the main reactor if required. A second solution would be to use recirculating reaction solvent in the condenser, cooling it with cold water via a heat exchanger. Synthetic oil might also be used as a condenser coolant, if compatible with the reaction mixture. A fourth solution would be replacing the glass condenser with a graphite block condenser, where the solid block ensures that leaks into the reactor are very unlikely. However, this would probably be the most expensive option.

No single book chapter can cover all the possible hazards that might be encountered on scale-up. Fortunately, the literature on plant safety is extensive. Indeed, there is an 'urban legend' of a scientist being seriously injured by a heavy safety tome falling on their head from an upper book shelf. Be that as it may, a good process chemist should keep abreast of the safety literature. Journals relevant to process safety include *Organic Process Research & Development* (OPRD), *Chemistry and Industry*, the *Journal of Loss Prevention in the Process Industries* (JLPPI) and *Process Safety and Environmental Protection*. However, incidents can be caused by poor communication rather than lack of knowledge. The importance of detailed, written shift-changeover notes cannot be underestimated. Another underlying cause of incidents is failure to follow standard procedures. Standard operating procedures (SOPs), batch documents and equipment manuals all play a vital role in ensuring plant safety and should not be ignored just because it seems convenient to do so. Mistakes in documentation do occur, but in most companies are comparatively rare, since SOPs and batch documents are checked and signed off by multiple individuals before being issued. Instructions are probably there for a good reason, even if you may not immediately be aware of it. If in doubt ask. Good communication is the greatest safety device.

Chapter 3
Purification

3.1 Measuring purity

It is one thing to have a process to make a compound and another to be able to consistently produce a product with acceptable purity. Firstly, you have to consider how the purities of the intermediates and the final product are going to be measured. This is the preserve of analytical development. In some companies, analytical development chemists are integrated into the process development department; in others, there is a separate analytical development department; and in some, there is a single analysis department encompassing both analytical development and quality control. All these three set-ups have their pros and cons, and it is not infrequent for senior management to order a change from one format to another.

The development of suitable analytical methods for a new compound and its intermediates is a challenging process, and there are many factors to consider, including selectivity, linearity, sensitivity and precision. It is unlikely that a method quickly produced by a development chemist one Friday afternoon will be the one that is finally chosen. In most cases, the analytical profiles for a final product and its intermediates are run by reversed-phase high-performance liquid chromatography (HPLC) or ultra-HPLC with UV detection. The percentage area value for the main peak from such runs is often referred to as 'the purity', although there are many reasons why it is unlikely to correspond to the actual purity. The response factors for the various peaks will probably differ, particularly if different chromophores are present, and there may be impurities present that aren't detected at all by UV. Other methods, such as HPLC with evaporative light scattering detection or GC, may be needed to analyse these. Common impurities not detected by UV include hexamethylsiloxane (Fig. 3.1 **a**) and hexaethylsiloxane (**b**), which tend to be formed from trimethylsilyl and triethylsilyl protecting groups, respectively. The latter compound is much less volatile, so is more likely to be a contaminant. Traces of long-chain aliphatics, either from oil pumps, oiled equipment or unintended coupling reactions, can also sometimes occur. The unintended reduction of aromatic rings, particularly some heterocyclic aromatics, during the reduction of other functional groups can lead to small amounts of unsaturated cyclic compounds that don't show up on UV. It is always best to use as mild a reducing system as possible for the task in hand. Charged compounds, including inorganic counter-ions in pharmaceutical salts [100], and many biomolecules, may be analysed by capillary electrophoresis or analytical ion chromatography.

The analysis of fragrance compounds is unusual, since their volatility and frequent lack of chromophores mean that GC is generally the first option for the final product and intermediates, rather than being merely reserved for impurities not detectable by HPLC. If GC is used as the main analytical method, additional methods

https://doi.org/10.1515/9783110717877-004

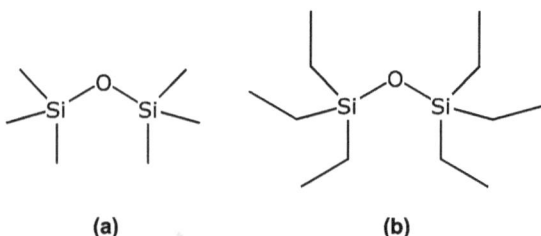

Fig. 3.1: Hexamethylsiloxane (**a**) and hexaethylsiloxane (**b**): impurities undetected by HPLC.

may be required to quantify non-volatile impurities or those that rapidly decompose on heating.

For generic compounds, an analytical method should normally be available. Most pharmaceuticals have standard methods in the US Pharmacopeia [101] or European Pharmacopoeia [102], although a few of these are rather old-fashioned. Many substances have published analytical methods in the literature, although not all will be suitable for routine industrial use.

Most final products are assayed against purified reference standards, either directly or by the use of a suitable purified internal standard. Major product impurities may also be assayed against reference standards, and the preparation of suitable standards is another job for the process development chemist (although, in some cases, they are commercially available). If product assays are consistently less than the % area of the product peak, it is probable that some unknown impurity is present, but isn't showing up on the chromatogram (it may not be detectable by UV detector, or it could be a so-called long-runner, which might appear in a subsequent chromatogram). This is not a situation that can be safely ignored.

In the pharmaceutical field, the modern 'quality by design' (QbD) approach to GMP manufacturing is now widely used [103]. The assay and impurity levels in pharmaceutical products are typically designated as a critical quality attributes (CQAs). Critical process parameters (CPPs) are those that affect CQAs, and have to be effectively controlled. CPPs may include process temperature, the addition of sufficient reagent, stirring rate, etc. The emphasis in QbD is on detailed understanding of reactions, rather than simply ensuring quality by testing. Outside the pharmaceutical field, such designations are less common, although nowadays similar control of output purity and process parameters is usually carried out.

It might be thought that products should be made as pure as possible. However, producing 'typical' batches of a novel compound with too high a degree of purity early on, particularly when preparing material for toxicity testing, can be a costly mistake if such purity levels cannot be replicated later. It is better to aim for realistic levels for hard-to-remove product impurities as a routine outcome for early batches, rather than have a very complicated and expensive process to get the level below 0.1%.

3.2 In-process controls and online monitoring

In-process controls are vital to consistently obtain the expected yields and purities in a plant. Such controls include checking the dryness of solvents, testing the pH of solutions, checking residual solvents during drying and tests to detect the consumption of the reactant, sometimes referred to as EOR tests (standing for either 'end or reaction' or 'end of run'). Tests can either be offline or online (sometimes referred to as 'inline' for continuous systems) and can involve various kinds of analytical technology. Offline EOR tests typically involve a sample being taken and analysed by HPLC, with a pass being given if the starting material is less than a particular value (say <0.5% area compared to the product peak). Such methods work best with relatively slow reactions where there is no danger of side reactions on prolonged stirring. In practice, TLC is quicker when speed is important, and with this technique there is no risk of delays due to HPLC instrument failure (but make sure you have a spare UV lamp). However, TLC will struggle to detect very low levels of starting material.

Online monitoring of reactions [104], sometimes referred to as process analytical technology (PAT), avoids sampling and chromatography delays. Near infrared (NIR) is a useful technique, with less interference from reaction mixtures compared to ordinary IR. The spectrometer can be placed at some distance from the reactor if a transmission probe and optical fibres are employed. Examples of reactions followed by NIR include a dehydration reaction carried out in THF in a continuous tubular reactor by Lundbeck chemists [105] and the bioconversion of guaiacol (Fig. 3.2 **c**) to tetraguaiacol (**d**) using horseradish peroxidase [106]. NIR has also been widely applied to formulation processes, such as tablet manufacture, as it is able to control various critical material attributes of such products [107]. Rather than simply looking at a single peak, in many cases chemometric techniques, such as multivariate analysis, are used for NIR calibration and measurement. Raman spectroscopy can also be employed for reaction monitoring in a manner similar to NIR, using optical fibres if required [108].

Oxidation or reduction reactions can sometimes be successfully controlled using redox electrodes, which measure the oxidation-reduction potential (ORP) of the reaction mixture. Here, you are typically looking at excess reagent at the end of a reaction, rather than the disappearance of the starting material. For example, a halogen or peroxide can be slowly added to react with the starting material, and there will be a clear change in the redox potential once the starting material is consumed and excess reagent is present. It may take some trial and error to find electrodes that withstand the reaction medium. Redox monitoring has also been used with peroxide-assisted enzymatic reactions [109]. In a similar manner, plant pH electrodes can be used to measure the completion of acidification or basification steps.

Benchtop NMR, incorporating suitable flow cells, can be used for online monitoring [110, 111]. This technique holds the promise of allowing the monitoring of unstable intermediates that would decompose on HPLC. NMR works best when the signals are

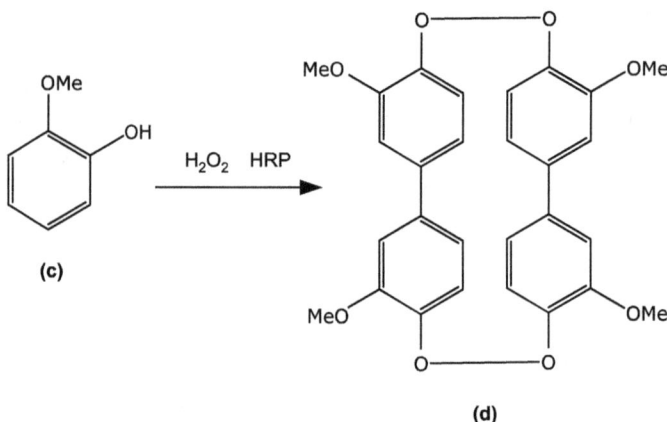

Fig. 3.2: Conversion of guaiacol (**c**) to tetraguaiacol (**d**) using horseradish peroxidase.

clear of solvent peaks. For instance, it is much easier to look for the disappearance of an aldehyde peak (which is typically way downfield of other peaks) than it is to follow changes to aliphatic signals that are close to large solvent peaks. In recent years, systems for online monitoring by mass spectrometry have also been developed [112]. Neither of these techniques is particularly common at present, but they may well become more prevalent in the future, since they offer the possibility of obtaining greater information about the chemical processes occurring in a reactor than other methods. Greater process understanding is an essential part of the modern QbD approach to manufacturing.

All online sampling methods can experience problems with fouling. The deposition of solids or sticky oils on the analytical device or sampling tube can lead to very misleading results. Online monitoring is easiest when there is little or no solid dispersed in the reaction mixture, and is particularly advantageous in continuous reactors, when rapid responses to changes in output quality may be required if acceptable purities are to be maintained.

3.3 Extraction and washes

The most straightforward way to remove impurities is by appropriate extractions and washes. Such steps avoid expensive chromatography or distillation, and are commonly used in a vast number of industrial processes. Ideally, several reactions can follow one after another using the same solvent, without any isolation of intermediates. For aqueous washes to be employed, a water-immiscible solvent must obviously be present. If the reaction is carried out in a solvent that is miscible with water, a second solvent will probably have to be added, while the initial solvent may have to be at least partially removed.

Separation by such washing is easiest when one component is acidic or basic and the other is not. For example, the oxidation of a primary alcohol may give a desired aldehyde, along with an unwanted carboxylic acid. The latter can be easily removed with bicarbonate or carbonate washes. Similarly, the reduction of a nitrile to an amine gives a product that will dissolve in aqueous acid, while any unreacted nitrile or amide intermediate will remain in the organic layer. Phenolic compounds can be advantageously dissolved in aqueous hydroxide solutions, allowing their easy separation from non-phenolic aromatics. Some 1,3-dicarbonyl compounds, sulphones and aliphatic nitro compounds can also dissolve in aqueous hydroxide, allowing for their separation.

Separation becomes tricky when two basic or two acidic compounds have to be separated, but even here, differences in pK_a values may make a separation possible. Generally, tertiary amines are more basic than secondary amines, which are more basic than primary amines. However, steric or resonance effects may change this order. The careful preparation of extraction buffers, such as mildly acidic phosphate buffers, may allow for successful separation [113]. Control of pH during the basification of the acidic extracts can also give further separation. Thus, a weakly basic amine might be liberated from its salt by the addition of bicarbonate solution, while a more strongly basic amine would remain largely as its salt, so would stay dissolved in aqueous solution.

Aldehydes and a few ketones can be removed from mixtures by forming a water-soluble bisulphite adduct (Fig. 3.3). Simply stirring with aqueous sodium bisulphite often gives poor conversion, but the rate of reaction can be hastened by the addition of some water-miscible solvent, such as methanol or DMF [114]. The non-aldehydic species can then be extracted with a water-immiscible solvent. Alternatively, in some cases, the adduct precipitates from the initial solution and can be filtered off. Regeneration of the aldehyde from the bisulphite adduct is accomplished with an aqueous base.

Fig. 3.3: Conversion of an aldehyde to its bisulphite adduct.

In some cases, non-polar products or impurities can be separated from more polar compounds by their extraction from polar solvents, such as methanol, DMF or DMSO, with an immiscible alkane solvent such as n-heptane. This method normally works only when there are large differences in polarity between the compounds to be separated. Counter-current liquid–liquid extraction methods can be employed in some cases to increase the efficiency of the extraction processes.

3.4 Distillation

Distillation processes are widely used in industry. The most common distillation is the removal of solvent from products. Just heating up a vessel under vacuum and condensing the solvent with a water- or glycol-cooled condenser is the simplest method of solvent stripping. However, it is a slow method, best used when a relatively small amount of solvent has to be removed (say, a small concentration of a solution so that a good yield of a solid product comes out on cooling); it is unsuitable for removing the large volumes of solvent that might be obtained from a dilute reaction or a preparative chromatography column.

On a relatively small scale, large rotary evaporators may be used, in a manner similar to laboratory solvent removal. However, the large flasks are awkward to handle and, in practice, a 50 L flask is the most you would want to deal with (even here, it is all too easy to accidently drop or crack the large flask). The capacity of rotary evaporators can be increased by feeding in solution to top up the flask.

Cyclic stills, also known as forced circulation evaporators, give fast evaporation of large quantities of solvent. The solvent is circulated around the apparatus by a pump, heated above its boiling point and then enters an evaporation chamber, where it partly vaporises and is carried over to the condensers. The exact configuration of the equipment varies from model to model. Cyclic stills are not suitable for the total removal of solvent, so typically the concentrated solution is removed once sufficient solvent has been taken off.

The removal of solvent is relatively straightforward, but the separation of reaction products by fractional distillation from by-products or starting materials is more challenging [115]. Distillation columns are typically needed to obtain adequate separation, and are available with a number of configurations – horizontal plates with 'bubble caps' being common. The ability of a column to separate compounds depends on the number of theoretical plates, which can be calculated for a given column. Your friendly chemical engineer may be glad to assist in such calculations. Modern software allows for useful theoretical models to be developed for fractional distillation. However, in practice, there may be a limited number of columns (maybe only one fairly short column) available on a particular plant; nobody will want to buy a taller one unless absolutely necessary. However, there is hope, since for any given column better separation can be achieved by increasing the reflux ratio. This is the ratio of the amount of liquid that returns to the column to the amount collected in the receiver. So, if 4 L are returned to the column for every 1 L that is taken off, the reflux ratio is 4:1. Higher reflux ratios give better separations, but at the cost of increased time.

Often, a high vacuum is needed to distil compounds, but this can be hard to obtain on plants, where there are so many places where air leaks can occur, and the vacuum pumps may be past their best. It is good to check the vacuum that can be actually obtained with a particular pilot plant or plant vessel before transferring a distillation process from the lab, where obtaining a high vacuum is much easier.

The fragrance industry is the go-to place if you want to see sophisticated distillations, with very tall distillation columns being widely used for separations, often of compounds with relatively similar boiling points. For example, the cyclisation of pseudo-ionone (Fig. 3.4 **e**) gives a mixture of α-ionone (**f**) and β-ionone (**g**), strong acids such as sulphuric acid favouring β-ionone, while weaker acids such as BF_3, phosphoric, formic or acetic acid favour α-ionone [116, 117]. α-Ionone has a distinct violet odour, making it a valuable perfumery ingredient, while β-ionone (still with a violet odour, but somewhat more 'fruity') is also used in perfumery, but finds is main use as a precursor to vitamin A and β-carotene. Vacuum distillation using tall columns and a high reflux ratio can separate the two isomers [118, 119].

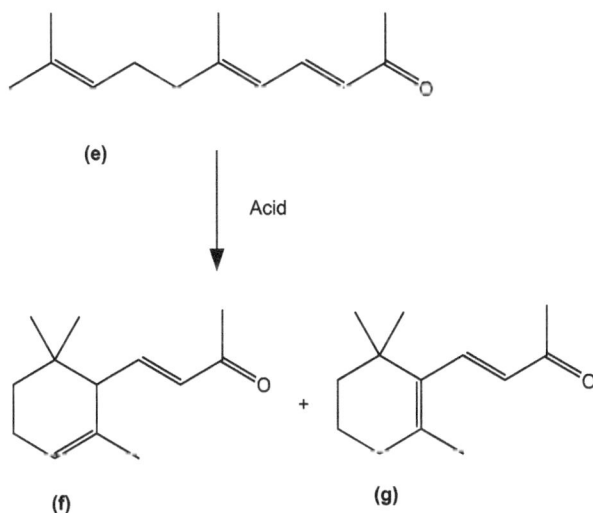

Fig. 3.4: Formation of α-ionone (**f**) and β-ionone (**g**) from pseudo-ionone (**e**).

Distillation becomes more complicated when three components have to be separated, i.e. light, medium and heavy fractions, which is difficult in a single column unless there are large differences between all three boiling ranges. An old-fashioned method is to have one distillation followed by another. The use of dividing wall columns (DWC, Fig. 3.5), in which a vertical wall is incorporated in the central portion of the column, allows a single column to give good separation of the three components, thus lowering capital cost and energy use [120]. A 'side-cut' ('middle-cut') is typically taken at the level of the wall centre, along with a 'top-cut' above the wall and a 'bottom-cut' below it. Adding extra walls can allow a greater number of components to be separated, although the design of such columns becomes quite complex.

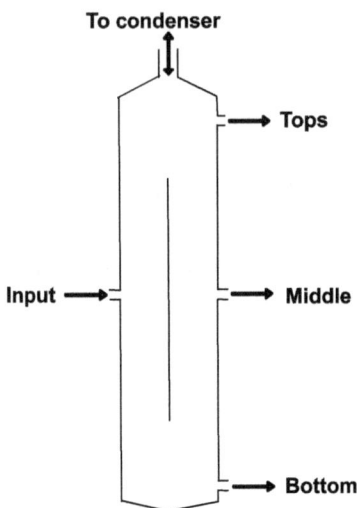

Fig 3.5: Simplified diagram of a dividing wall column (horizontal plates and packing not shown).

3.5 Crystallisation

Many final products are crystalline solids, either salts or neutral compounds. Crystallising a solid from solution gives an opportunity to leave unwanted by-products and the starting material behind, while a subsequent recrystallisation can further improve a product's purity. Although taking an intermediate in a solution from one stage to another reduces the processing time and maximises yield, too many impurities can build up if such a procedure is carried out over too many steps. For the majority of longer syntheses, it is best to crystallise and isolate at least one of the intermediates. However, this may be easier said than done in cases where most of the intermediates are oils, such as the prostaglandin syntheses described in chapter one. As mentioned previously, the Chirotech synthesis of travoprost [14] has the advantage that one of the key intermediates is crystalline. Generally, in such cases, it's worth spending some time screening a variety of solvents to find at least one intermediate that can be crystallised. In the conventional travoprost synthesis, the diol compound (Fig. 3.6) is a low-melting solid, while the other intermediates are oils.

Ideally, a reaction product would crystallise from the reaction solvent merely on cooling, but, in practice, it is common for either all of the material or a large percentage of it to remain in solution. Thus an antisolvent, in which the solid has very low solubility, is often required. For example, if a reaction is carried out in

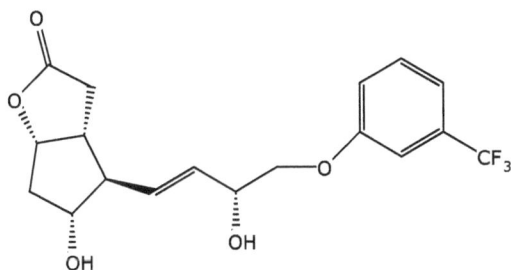

Fig. 3.6: Solid diol intermediate in travoprost synthesis.

isopropyl acetate, n-heptane might be added to the hot solution, giving a much better yield of solid on cooling than could be obtained in its absence. You want sufficient antisolvent to give a good yield, but not so much that the solid does not dissolve in the hot mixture. Crashing out the solid by adding too large a quantity of antisolvent normally crashes out many impurities. Whether added hot or cold, rapid mixing of the antisolvent is normally essential in order to obtain a uniform product.

In some cases, a complete solvent swap is needed to obtain crystals, which is easy enough if you're going from a low-boiling to a high-boiling solvent, but very difficult the other way round, unless a suitable azeotrope is formed. Useful solvent swaps for plant processes have been described in the literature [121].

It is important to optimise the crystallisation conditions so as to control the level of impurities. Impurity levels in active pharmaceutical ingredients (APIs) are, typically, CQAs, thus making the crystallisation condition CPPs. Impurities can contaminate crystals due to crystal agglomeration, inclusion within crystals, surface deposition or co-crystallisation. Microscopy studies can give valuable information about which of these processes is the main cause of low-purity crystals, and appropriate measures can be taken to combat it. In the case of surface deposition, extra washes may well sort out the problem, or else producing larger crystals to allow for better removal of the mother liquor. Agglomeration may be avoided by having a lesser degree of super-saturation before crystallisation or by changing the solvent entirely. A useful structured approach to reduce impurity levels has been produced by a team from the University of Strathclyde, Bayer, Pfizer and Eli Lilly [122].

Mean particle size is an important parameter, since too fine a precipitate will take an unacceptably long time to filter, sometimes stretching to days. Too high an agitator speed can break up crystals, leading to an increase in the proportion of fine particles and consequently slow filtration. In the case of final products, the particle size distribution often affects the formulation, so may be classed as a CQA. In such cases, very careful control of the crystallisation parameters, including any seeding procedure, is necessary in order to control the crystal output within an acceptable range. If seeds are added to a solution that is not saturated with product, there is a danger that they will dissolve, and if the degree of supersaturation is too high, crystallisation may occur before seeding, which in some cases may lead to less reproducibility. Instrumentation, such as focused beam reflectance measurement (FBRM) probes [123, 124], which are based on the reflectance of laser beams, and near-IR probes [125], can be used to follow the crystallisation process within the reactor.

A variety of industrial filtration units are available for filtering crystals. These include ceramic filtration units (typically used with an appropriate 'filter cloth') and metal Nutsche filter-dryers. Care should be taken to ensure that the construction materials are compatible with the reaction mixtures. Although filtration is the normal means of isolating crystals and other solids, centrifuges are also employed, particularly for more amorphous materials and biological samples.

Incorporating crystallisations into continuous processes presents its own challenges [126, 127, 128]. In theory, continuous crystallisation should allow for better control and reproducibility, but blockages can be a problem in practice. Continuous crystallisation can be carried out in a standard vessel with an overflow outlet, so that the inflow of solution (and antisolvent, if required) matches the outflow of slurry. Various types of purpose-built crystallisation equipment can also be used, such as a coiled-flow inverter (CFI) crystalliser [129]. Here, crystallisation takes place in a helical tube, which has 90° bends equally spaced along the length of the helix.

One special area of crystallisation is its use to separate enantiomers, usually by the formation of diastereomeric salts with suitable chiral acids or bases. Such processes are used less often than in the past, due to the wide range of stereoselective reactions now available, but are still required for the production of some compounds [130, 131]. Chiral acids, such as tartaric acid or its derivatives [132], are commonly used for the resolution of amines, while racemic carboxylic acids can be resolved by the use of chiral bases or amino acids, such as S-lysine [133]. Resolution of racemates by diastereomeric crystallisation has a maximum theoretical yield of 50%, but, in practice, yields are significantly lower, wasting large amounts of valuable material. However, the situation can be salvaged in some cases by the racemisation and recycling of the unwanted isomer. For example, the racemate of the weight loss drug lorcaserin (Fig. 3.7 **h**) was reacted with L-tartaric acid to give the tartrate salt of the desired R-isomer (**i**), which was isolated by filtration [134]. Treatment of material derived from the filtrate, predominantly the unwanted S-isomer (**j**), with a strong base, such as potassium t-butoxide in DMSO, regenerated the racemate. This is an unusual racemisation, since the chiral centre is at a β-carbon with regard to the heteroatom.

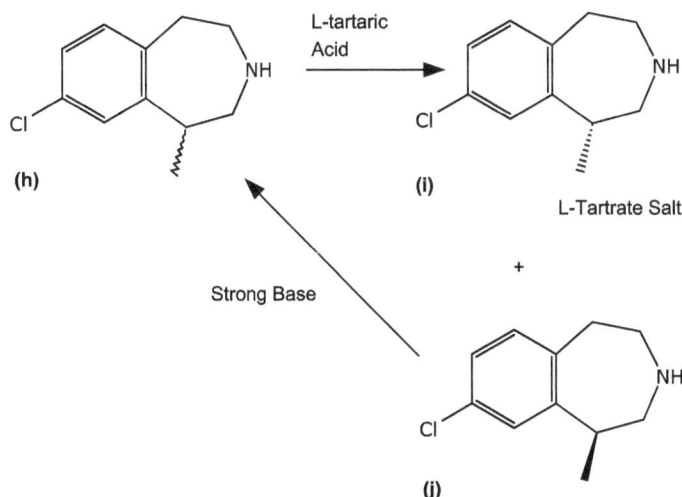

Fig. 3.7: Diastereomeric resolution of racemic lorcaserin (**h**), with racemisation of unwanted isomer (**j**).

3.6 Polish filtration and impurity removal

Small amounts of various solids suspended in product solutions can contaminate the product while also favouring uncontrolled crystal nucleation, rather than controlled nucleation from the added seeds. So-called 'polish filtration' is used to remove unwanted solids, where the bulk solution is sucked or pumped through a membrane, 'filter cloth', filter paper or filter cartridge, the latter typically being used when only very small quantities of solid are present, since they can be easily blocked. Various filter aids, such as Celite®, Celatom® and Florisil®, are often employed in difficult cases, being used to form a bed on top of the 'filter cloth'. The former two solids are diatomaceous earths, while the latter is a synthetic magnesium silicate.

Sometimes, your unwanted impurities are in solution, but all is not lost. If your product is non-polar and the impurities very polar, simple filtration through a bed of silica or alumina may remove the unwanted substances. Such processes can be particularly effective with brown tarry material from oxidation reactions, degraded phenols, etc. Filtration through silica or alumina beds is actually a crude form of chromatography, and sufficient washes need to be employed to ensure that none of the desired product remains on the solid. It is normally safest to prepare the bed on the filter by using a slurry with a fresh solvent, rather than adding the process solution to dry silica, since channelling is more likely in the latter case.

Activated charcoal can be employed in a similar fashion to remove coloured material, although in this case a pre-slurry in a vessel with the process solution, followed by filtration through a filter aid, may be more effective than simply filtering through a bed of charcoal. For example, a recent patent application describes a recrystallisation of the fibrosis drug pirfenidone, which involves stirring a hot solution with charcoal for 1 h, before removing the charcoal by filtration, followed by cooling to give the product [135]. It's worth examining different grades of charcoal to find the one that works best with your process.

3.7 Drying

The final purification stage for many compounds is drying. The simplest way of drying is to spread the product on trays in an oven, which works fine for many compounds. Often, a flow of nitrogen is used to remove solvent or water, while preventing any oxidation, or a vacuum oven may be employed. However, many compounds will be in the form of large lumps when initially placed in the oven, and these will have to be broken up to ensure even drying and a suitable product. The result is many hours of 'lump-crunching' in an air-fed suit, the process becoming more time-consuming and expensive in terms of labour costs the larger the batch. Commercial lump breaker machines are available to speed up the process.

Combined Nutsche filter-dryers make both filtration and drying straightforward and minimise operator exposure, but are not suitable for all compounds and mother liquors. For example, a stainless steel filter dryer will be incompatible with strong acids or high chloride-ion concentrations. Filter dryers made from Hastelloy® alloy allow for a greater range of mixtures to be filtered and dried, but are significantly more expensive.

For unstable compounds, freeze-drying (lyophilisation), is sometimes used. This works best with water but can also be used to remove some organic solvents, such as DMSO [136]. Spray-drying is also used to produce dry compounds or formulations from concentrated solutions or slurries [137].

3.8 Troublesome impurities

There are some impurities that are a nightmare for process chemists, and dealing with them can take up a great deal of time. Plasticisers, such as di-n-butyl phthalate (DBP, Fig. 3.8), can contaminate products. Often, it takes a degree of detective work to determine the source of such compounds. Solvents are a possible culprit, particularly if they have been transferred from their original drums into secondary containers from which they can pick up traces of plasticisers. Contamination is particularly likely when large volumes of solvent are used for preparative chromatography. There are many materials that contain plasticisers, and it may be that something such as an old nylon gasket on the plant is the source of the problems. Contamination may also occur in the analytical

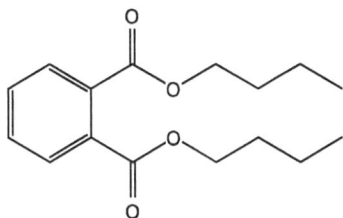

Fig. 3.8: DBP, one of a number of plasticisers that can contaminate products.

lab, and it is a good idea to run a variety of blank samples, including some sampled in the same manner as plant samples, to rule this out.

Other awkward impurities are by-products from reactions. Triphenylphosphine oxide (TPPO), a by-product from the use of triphenylphosphine in Wittig or Mitsunobu reactions, is notoriously difficult to completely remove. It has some solubility in most solvents, and tends to 'spread out' on preparative chromatography, so it is often not separable from the product by this technique. Polymer-supported triphenylphosphine has been widely used [138] and gives an oxide by-product that is easily removed by filtration. However, the polymer-supported reagent is more expensive and in some cases gives lower yields than triphenylphosphine.

With regard to the Wittig reaction, the Horner–Wadsworth Emmons (HWE) reaction should be used, if possible, since the phosphate by-product is water soluble and easily separated. The HWE reaction normally gives good yields of *trans* olefins from stabilised phosphonate anions, but careful control of the reaction conditions is needed to obtain significant amounts of *cis* olefins [139]. A recent new Mitsunobu method [140] from Ross Denton's group at the University of Nottingham uses a catalytic amount of a phosphorus oxide-containing catalyst (Fig. 3.9) rather than stoichiometric amounts of triphenylphosphine, and it seems to be a very promising alternative to the traditional reaction.

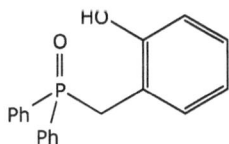

Fig. 3.9: Phosphorus oxide-containing catalyst used in new Mitsunobu method.

Dicyclohexylcarbodiimide (DCC, Fig. 3.10 **k**) is used in a variety of reactions, including peptide coupling [141] and other reactions of carboxylic acids, such as amide and ester formation. The by-product, dicyclohexylurea (DCU), has fairly low solubility in many organic solvents, and the resulting precipitate can be filtered off at the end of the reaction (although, some residual material can still remain to potentially contaminate products). DCC is not always suitable when solid-phase synthesis methods are used, since DCU can precipitate within the solid phase. DCC also has a bad reputation for giving rise to allergic reactions, so it is best avoided on plants. Alternative diimides, such as diisopropylcarbodiimide (DIC, **l**), give ureas greater solubility in organic solvents, while the urea from 1-ethyl-3-(3-dimethylaminopropyl)carbodiimide (EDAC or EDC, **m**) is water soluble. Whichever diimide is chosen, it is advisable to check for residual amounts of the respective urea and any unreacted diimide in subsequent intermediates and products.

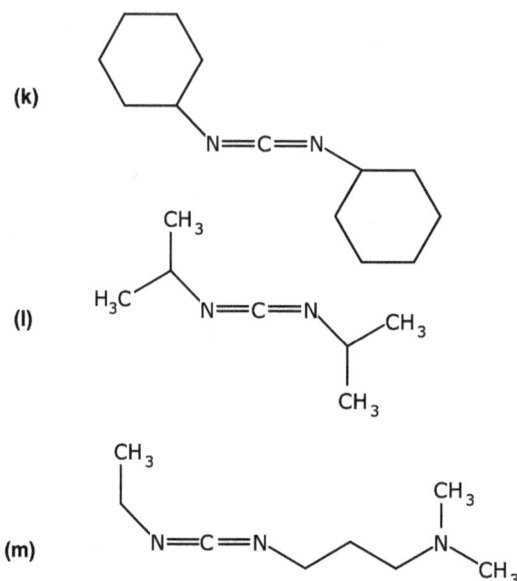

Fig. 3.10: Diimides used for activating carboxylic acids: DCC (**k**), DIC (**l**) and EDAC (**m**).

Removing traces of heavy metals can be a challenge, and metal scavengers [142, 143] may be needed. These are discussed in more detail in Chapter 7. Mutagenic impurities, which have become a topic of great concern over recent decades, are also discussed in that chapter.

3.9 Preparative chromatography columns

In laboratory chemistry, flash silica columns are often used, being the obvious method of purification. In the case of industrial processes, there is one answer to the question, 'Should we run it down a column?' and that is 'Don't'. At any rate, do everything in your power to purify intermediates and products by washes, distillation, recrystallisation, etc. Preparative columns on a large scale are time-consuming and expensive. As far as agrochemicals are concerned, it is often said that any product that requires preparative chromatography will be too expensive to be spread on the fields. There are some cases where this rule of thumb may not always hold true, such as pheromones for trapping devices and other bio-regulators [144], but, in general, it holds good. In practice, preparative chromatography is mainly used for relatively high-value pharmaceuticals.

In some cases, home-packed flash columns can be used for preparative chromatography, in a similar manner to laboratory columns, except that they may stretch from floor to ceiling. Even so, a tall 2 m column separating two compounds of fairly similar polarity might only be able to cope with an input of say 1 kg, meaning 20 runs would be needed for 20 kg of input. Columns are normally run under nitrogen pressure, and relief valves are needed to prevent over-pressurisation. Larger columns are often made from stainless steel rather than glass, allowing for greater pressures. If glass columns are used, plastic coating or mesh reduces the risk from glass fragments, in the event of over-pressurisation causing the walls to crack. Careful consideration is needed with regard to the choice of solvents, good separations, reasonably low toxicity and easy solvent stripping being required. The product spot should be near the bottom of a TLC plate of the proposed solvent mixture, but separated from impurity spots. In practice, gradient elution is usually employed, often starting with a solvent mixture having the minimum polarity needed to dissolve the input.

Fractions are typically examined by a UV detector or TLC, followed by analytical HPLC (run on a concentrated sample taken up in the HPLC solvent). It can save analytical runs if suitable combinations of fractions are combined for analytical HPLC, rather than submitting each individual fraction. Thus, equal amounts of (for example) fractions 15–23 are removed with an adjustable-volume pipette, combined together, concentrated on the rotary evaporator and submitted for analytical HPLC.

Home-packed columns generally employ 'flash' silica (40–63 µm particle size or similar), but alumina or charcoal can also be used when necessary. The latter is useful for separating some phenolic compounds and natural products [145], but often many different types of charcoal need to be examined in order to find one that gives good separation. Ion exchange resins can be used for charged molecules, such as amino acids, peptides and proteins [146, 147].

Silica gel columns are best packed as a stirred slurry rather than dry packed, since overheating or cracking can occur if the latter method is used. Cracking and channelling are likely to occur if the column is ever left to run dry, so care must be taken to pump in solvent regularly. Such a mishap is particularly liable to occur with large stainless steel columns, where the solvent level is not readily visible. You may spend a lot of time peering through the sight glass at the top of the column to make sure the solvent level is correct. Since a large number of columns are needed to produce a significant amount of product, the expense of silica becomes quite significant, and it is not feasible to throw the silica away after each column. Typically, the used column is flushed with a reasonably polar solvent, and then eluted with the starting solvent. Thus, a gradient column, starting with 4:1 n-heptane: ethyl acetate, might go through 3:1 n-heptane: ethyl acetate and 2:1 n-heptane: ethyl acetate.

If the product is eluted at this point, the column would be washed with neat ethyl acetate to remove polar impurities, and then washed again with the 4:1 n-heptane: ethyl acetate prior to the next run. However, such a sequence might fail to elute very polar impurities and might also lead to the build-up of tarry material at the top of the column. Putting through a very polar solvent, such as methanol, to clean the column may impair the subsequent column performance.

Figure 3.11 shows a typical set-up for a home-packed flash column. It is important to place sand below the silica (to prevent blockages) and above (to preserve the top surface of the silica). It is best to pump in the solvent via a spray ball; otherwise, a crater may form if the flow is directed at the column surface or at one point on the walls.

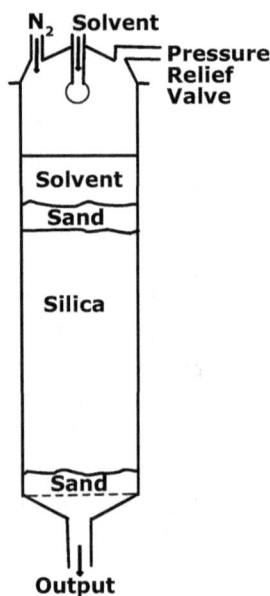

Fig. 3.11: Home-packed flash column with spray ball solvent inlet.

Popular alternatives to home-packed columns for flash chromatography are commercial pre-packed columns and the associated systems [148], which are produced by companies such as Biotage and Teledyne ISCO. Normal-phase, reversed-phase and ion exchange columns are available, packed in plastic cartridges, and these generally give better separation than home-made columns for the same amount of silica. Often, a somewhat smaller grade of silica is used compared to the 40–63 µm particle size typical for home-packed columns. A wide variety of column sizes are available, allowing for gradual scale-up and reaching a multi-kilogram input scale for the largest systems. The pre-packed columns allow for more runs to be carried out than home-packed columns before their performance becomes unacceptable, so they tend to work out cheaper in the long run, despite the high initial cost of the equipment and cartridges.

An alternative to flash chromatography is simulated moving bed (SMB) chromatography [148, 149, 150], with systems offered by companies such as Knauer, Novasep and XPure. With SMB chromatography, a number of columns, typically eight, are connected in series in a circular manner. The aim is to obtain a similar result as would be achieved if the solid column packing moved in a counter-current fashion to the eluent. Of course, the solid does not actually move, but an equivalent result can be obtained by switching the input and output to the various columns. A number of variations on the initial SMB concept have been introduced over the years. SMB can be carried out continuously on a large scale, producing tonnes of product if necessary. Normal phase, reversed phase and resins can all be used, and the method has also been applied to the separation of enantiomers [151, 152]. It uses relatively less solvent than flash chromatography, making it suitable for large-scale production. SMB works best when separating two components and can struggle when the product peak elutes between two impurities.

The main use of preparative HPLC in process development departments is the isolation of impurities for characterisation. However, it is used as a process purification method for high-value low-volume products, such as prostaglandins [13]. Since an output per run of a few grams, at most, is typical with standard columns (5–20 cm diameter), many runs have to be carried out to even make a kilogram of product. Specialist industrial-scale HPLC columns and associated equipment are available from companies such as Novasep (with diameters up to 1 m or so), but are costly. As a variety of chiral columns are available, preparative HPLC can be used to produce pure enantiomers when other separation methods fail [153]. Sometimes preparative HPLC is used to produce a single enantiomer of a drug for clinical trials, even if the process is not really suitable for full-scale production. The argument being that since most drugs fail clinical trials, chances are that a large-scale method will never need to be developed. However, the process chemist is then faced with the need to rapidly devise and develop a new industrial process to give the required enantiomer, should the clinical trials actually be successful. Telling senior management that you gambled on failure may not be the best career move.

Regardless of the type of column employed, vast amounts of solvent will be produced, and full-scale production will normally require effective solvent recycling. Since mixed gradient solvents are often used, distillation can produce different distillate fractions of varying composition. Accurate analysis by GC is needed to determine the amount of each solvent in the various drums of distillate, with careful mixing and matching to reconstitute the correct solvent ratios for re-use. For gradient reversed-phase chromatography, a methanol/water system allows for easier solvent recycling than acetonitrile/water, since the boiling points are relatively far apart in the former case and no azeotrope is formed. Overall, it is not just the cost of solvent, silica and equipment that makes preparative columns expensive on a large scale, it is also the extra analytical support needed for the seemingly endless fraction analysis and solvent recycling.

One separation method that uses lower amounts of organic solvent is supercritical fluid chromatography (SFC). Here, the bulk of the solvent is supercritical carbon dioxide, while a smaller amount of a polar solvent, such as methanol, is added to give a gradient to remove the product from the column. SFC has been applied on a preparative scale and can be considered a 'greener' preparative chromatography method, due to its low solvent usage. Like preparative HPLC, it can be used to separate enantiomers [154]. SFC gives speedy product elution, so a larger number of runs can be carried out per shift than with a comparable preparative HPLC system.

The performance of any column can be measured in terms of theoretical plates, in an analogy to distillation columns [155, 156]. Of course, there are no actual plates present, but such hypothetical calculations allow for a comparison of different systems and chromatographic methods. However, with preparative chromatography, higher numbers of plates are not necessarily useful if they are simply obtained by reducing the particle size. Columns using a very fine particle size overload very quickly and require very high pressures, so are not usually favoured. Hence, the widespread use of 40–63 μm (or 20–40 μm) particle size silica for flash columns.

There is a given optimum flow speed for any given column. With too low a flow, diffusion of solutes becomes increasingly significant, while with too high a flow, the solutes are too far from equilibrium with the solid phase to give optimum separation. The van Deemter equation [157, 158] relates theoretical peak height (H) to the linear velocity (u) of the mobile phase:

$$H = A + (B/u) + (C \times u)$$

A is a term reflecting eddy diffusion, B is a term reflecting longitudinal diffusion and C is a term reflecting the resistance to mass transfer. The use of this equation shows a clear minimum for H, where separation is optimum. Figure 3.12 shows a typical van Deemter curve for a column. The use of the van Deemter equation has been criticised and modifications proposed [159], although it is clear that there will be an optimum flow speed for each column, regardless of the exact mathematical details.

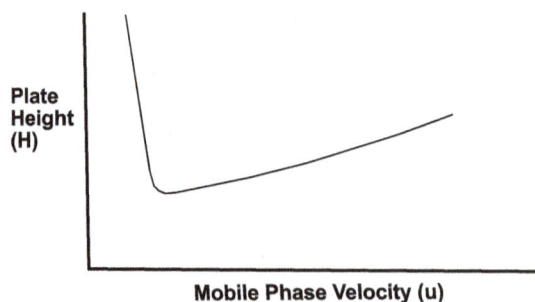

Fig. 3.12: Hypothetical van Deemter curve for a chromatographic column.

Although there are many different chromatographic separation methods, most small-molecule production processes use either flash chromatography or SMB systems since these are most amenable to rapid scale-up at a reasonable cost. However, technological innovations are continually occurring, so other techniques may become more popular in the future.

Chapter 4
Polymorphs

4.1 Analysis, formation and screening of polymorphs

Sometimes when process development seems to be complete and full-scale manufacturing may have been carried out for a while, the production process can start to go disastrously wrong. The formulated product may suddenly crash out of solution, or routine tablet formation may fail. Such catastrophes are both costly and time-consuming to remedy. Patients may be deprived of vital drugs while the problems are being sorted out. The root cause is that one polymorph (crystalline form) is being produced rather than another, and this can have profound effects on filtration, formulation and even bio-availability [160].

The majority of crystalline solids can form more than one polymorph, while an amorphous (non-crystalline) form may also be formed in some instances. Polymorphs differ in their molecular packing, so differences between them can be revealed by X-ray methods, such as single crystal diffraction or X-ray powder diffraction (XRPD). Other techniques, such as attenuated total reflectance Fourier transform infrared spectroscopy (ATR-FTIR), Raman spectroscopy, solid-state NMR and differential scanning calorimetry (DSC) can also be used to distinguish between different polymorphs, although in some cases the differences may be slight. XRPD [161], solid-state NMR, Raman and ATR-FTIR [162] can be used to quantify the amount of polymorphs present where mixtures occur. Not all these techniques will be available in the average analytical laboratory, although various academic groups and industrial consultants are able to provide a routine service if necessary.

A particular polymorph will be more thermodynamically stable than another at a particular temperature and pressure, and phase diagrams can be drawn showing these differences. For example, a 2017 paper [163] describes phase diagrams for metacetamol, the *meta* analogue of paracetamol (acetaminophen) (Fig.4.1 **a**), which exists in different polymorphic forms, as does paracetamol (**b**) itself [164].

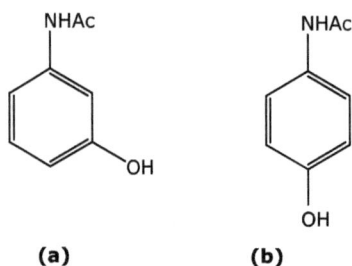

(a)   **(b)**

Fig. 4.1: Metacetamol (**a**) and paracetamol (**b**).

https://doi.org/10.1515/9783110717877-005

However, although phase diagrams tell us which isomer is stable under different conditions, they tell us nothing about the rate of transformation. Diamond is thermodynamically somewhat less stable than graphite at room temperature and atmospheric pressure, but diamonds do not suddenly disappear into a heap of soot. In general, the stronger the intermolecular interactions, the slower the transition from the metastable to stable polymorph. So if the polymorphic form you are using seems perfectly stable, this may be simply due to its imperceptible transformation to the more stable form.

The danger with polymorphs is that companies may work for years with a form that they believe to be stable, but once the more stable polymorph is formed (maybe by some perturbation from the usual conditions), it can be virtually impossible to produce the initial polymorph, even under conditions that reliably gave rise to it prior to the formation of the new polymorph. The original form is now a 'disappearing polymorph' [165, 166], only surviving in diminishing stocks in the freezer. On occasions, attempts have been made to transfer the crystallisation to other sites, or even cross oceans, in an effort to avoid any seeding from the unwanted form. Such attempts can be successful, particularly if new equipment and staff are employed. On other occasions, attempted preparations at new sites fail to form the original polymorph, and there seems to be an almost occult jinx on all such endeavours. However, there is no need to resort to supernatural explanations: any bulk solid will have vast numbers of tiny particles of varying sizes floating in the air above it, many of which will be too small to be seen even with an optical microscope. These potential seeds rapidly spread into the atmosphere and onto surfaces. Just as pollutants and microplastics can arrive at remote mountain locations, so particles are blown around the globe. In addition, staff move from site to site, and they can bring such particles with them. Clothing, hair, cars and paperwork can all move sub-microscopic seeds to a supposedly 'uncontaminated' site, and even if the chemists are scrupulous about avoiding any such opportunities for transfer, what about the financial staff, the visiting sales rep or the factory inspector? Today's industry is both interconnected and international – no site is an island.

To avoid unpleasant surprises, it is important to thoroughly screen for potential polymorphs. High-throughput crystallisation runs involving a diverse array of solvents are a useful method for obtaining novel isomers [167]. The formation of solid without solvent, i.e. from solidified melt, also needs to be examined. Of course, if you have a list of, say, 100 solvents, it is quite likely that the product will have little or no solubility in some of them, even on heating. So co-solvents will be needed. Using solvent mixtures in addition to neat solvents and varying the temperature and concentrations lead to a requirement to carry out thousands of runs. Varying degrees of automation can be used to make the screening process less tedious – such as adding solvents to either multi-well plates or multiple glass tubes.

Recrystallisation in the presence of various additives, particularly, compounds that act as templates, may help produce new polymorphs [168]. A wide variety of templates have been used, including inorganic solids, polymers and organic solids. A good 'fit', referred to as epitaxy, between the structure of the template surface and the growing polymorph crystals, favours the formation of the latter by a process of regular layer-by-layer growth. In addition to epitaxial factors, molecular interactions such as hydrogen bonding between the template and developing polymorph may also play a crucial role. Polymorphs can be formed by confinement in suitable nanomaterials, such as polymer networks or gels. Control of pore size and shape in template materials such as polymers or glasses can change the polymorphic outcome. For example, paracetamol (acetaminophen) transitions have been examined in nanoporous glasses and related to the pore size [169]. Polymer microgels, which can have tuneable structures, have been shown to control the crystallisation of various polymorphs [170]. The crystallisation of mixtures of thalidomide (Fig. 4.2, **c**) and barbital (**d**) was controlled by specially synthesised urea gels (produced from compounds such as **e**), with structures mimicking those of the drugs [171].

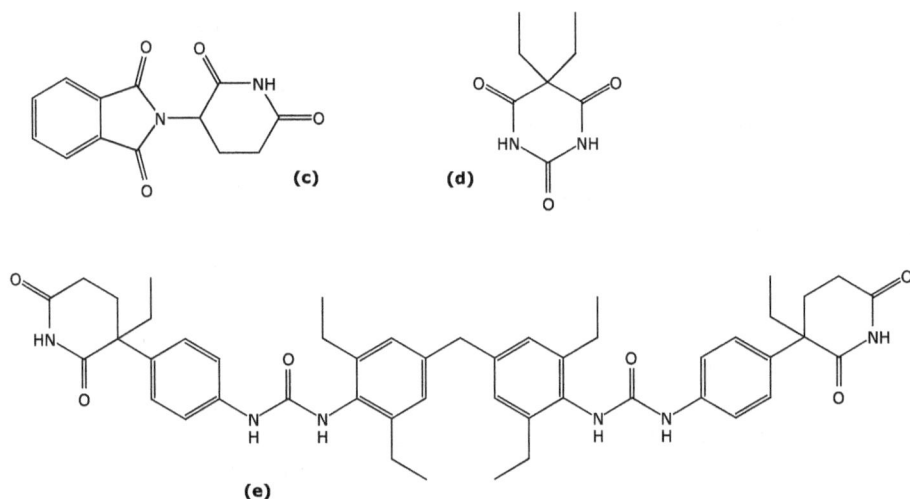

Fig. 4.2: Thalidomide (**c**) and barbital (**d**) and an organogelator compound (**e**).

Novel polymorphs may be produced by high-pressure techniques [172]. For example, the drug piracetam (Fig. 4.3, **f**) was converted from Form II to a novel form (Form V) by compression at 0.45–0.7 GPa [173]. A 2016 paper [174] described extensive investigations on the polymorphic transformations of the diabetes drug chlorpropamide (**g**) under pressure. Surprisingly, different polymeric forms were produced using different inert pressure-transmitting media (helium, neon or paraffin), despite chlorpropamide being insoluble in these substances.

Such high pressure transformations require specialist equipment, although simple grinding with a ball mill can give rise to polymorphic transformations. The results can differ depending on whether grinding is carried out in the presence or absence of solvent [175].

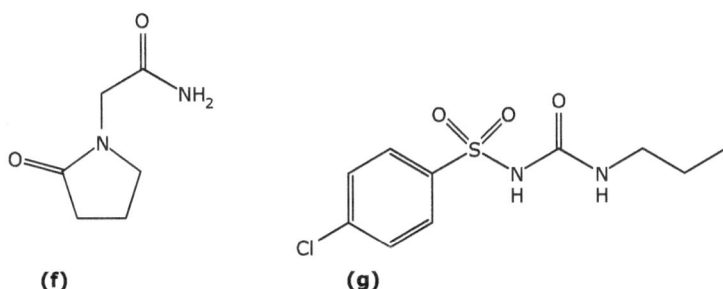

Fig. 4.3: Piracetam (**f**) and chlorpropamide (**g**): Both undergo polymorphic change at high pressure.

Sublimation under reduced pressure is another way to produce polymorphs, often giving different results to recrystallisations. The formation of a particular polymorph can be favoured by sublimation onto particular templates [176, 177]. In some instances, the success of sublimation in producing new polymorphs has been linked to structure, such as the hydrophilicity of amino acid side chains [178].

Thus, there are a wide variety of methods for forming polymorphs, and in the majority of cases you will eventually find more than one form of a crystalline solid if a thorough investigation is carried out. You can then determine their relative stability at different temperatures, although this may prove difficult if kinetic factors make the transformation of one form to another imperceptible. On the whole, the most stable form will have the highest melting point.

As far as pharmaceuticals are concerned, vital properties of the drug substance, such as particle size distribution and percentage of the correct polymorphic form present, will usually be 'critical quality attributes'. The understanding and control of these is part of the overall 'quality by design' process. Regulators will want to ensure that the factors affecting polymorphic composition are well understood; nowadays, simply testing for compliance is insufficient on its own.

4.2 Polymorph patents

If you are introducing a new compound to the market, obviously you need to patent it. You'll also want to patent any polymorphs, particularly the form you wish to market. A polymorph patent can be used to extend the effective protection for a compound. If you first of all patent the compound, and the polymorph described or implied is not suitable for formulation, you can then patent the useful polymorph, say 3 or 4 years later, creating a headache for generic competitors, who would not be able to use the useful form when the first patent expires. However, if you leave too long a gap before protecting the useful form, there is an increasing risk that another company will study your compound and patent it themselves.

The ease of patenting polymorphs varies depending on jurisdiction [179]. Full disclosure of the polymorph preparation method and characterisation data will be required. As far as European patents go, an inventive step is necessary [180]. In practice, a polymorph must have unexpectedly superior properties to known forms, such as stability and bio-availability. If no polymorphs are described previously, it may be hard to patent your preferred form, since it can be argued that a scientist 'skilled in the art' is inevitably going to search for the most stable or most bio-available form (i.e. a search for a stable polymorph is not classed as inventive, under these circumstances). The US courts have proved somewhat more likely to uphold US polymorph patents from challenges on the grounds of 'obviousness'. In India, polymorphs, solvates and salts are all considered to be the 'same substance' as the original patented substance, unless a clear improvement in efficiency can be demonstrated [181]. Therefore, properties such as stability or filterability would not be grounds for patenting a new polymorph of a compound. Such limits on patentability make it harder to use polymorph patents to extend patent protection for drug substances, allowing generic suppliers to access the market sooner rather than later, and thus lowering drug prices.

Since patent examiners and courts in different jurisdictions can reach different conclusions, it is quite possible for a polymorph patent to be upheld in one country but struck down in another. It is obvious that the patenting of polymorphs is an extremely complex business, and high-quality specialist legal advice should always be sought when drafting patents or attempting to circumvent them. Getting it wrong can be extremely costly to your company.

In addition to polymorphs, novel salts, solvates and even amorphous forms can potentially be patented. Both salts and solvates may exist in more than one polymorphic form, adding to the complexity of the patent landscape. For example, a Sandoz patent [182] claims a novel acetone solvate of the heart drug ivabradine hydrochloride (Fig. 4.4), which is useful as a precursor to final drug substance, giving the correct polymorph that is free from acetonitrile. Previous methods had involved crystallisation from acetonitrile, which led to excessive levels of this toxic solvent being incorporated into the product.

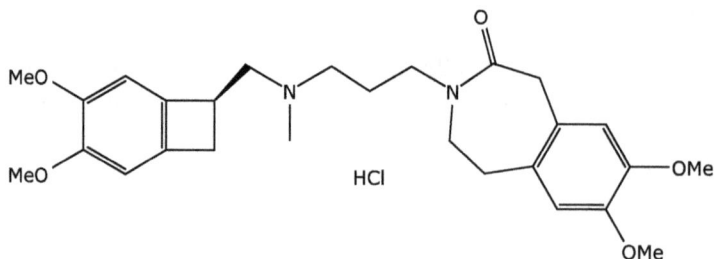

Fig. 4.4: Ivabradine hydrochloride, which can be produced via its acetone solvate.

If you are working for a generic company that is planning to produce a substance on which the patent is about to expire, the polymorph patent landscape can be daunting. In some cases, a number of companies will have claimed various polymorphs, while others may be disclosed in literature articles. For example, the initial patent from 'pharma company A' may mention one solid, while a patent 2 years later may refer to the initial form as Form I, but claim a novel polymorph, Form II, which is stated to be more stable and suitable for tablet formation. Some years later, 'generic company B' files a patent claiming three new polymorphs, named Form III, Form IV and Form V. The patent is disputed, with 'pharma company A' claiming Form III is really just impure Form I, while Forms IV and V are actually solvates and unsuitable for production. Meanwhile, 'generic company C' claims Form Z, with an XRPD spectrum which is rather similar, but not quite the same, as Form II. Later, academic group D describes Form Y in a paper, with a different IR spectra to Forms I, II and Z, but with no XRPD provided. Thus now you face a landscape of mist and confusion, which will take a lot of patient work, both scientific and legal, to clarify.

4.3 Case study 1 – ritonavir

Ritonavir (Fig. 4.5) is a protease inhibitor from Abbott used to treat HIV/AIDS, nowadays in combination with other drugs. It is a peptide mimic containing four chiral centres. The initial process development of the drug seemed to be successful; manufacturing had started in 1996 and continued smoothly for 2 years before problems emerged. The drug was formulated as semi-solid capsules containing the active ingredient in a solution of aqueous ethanol, since the drug lacked bioavailability in the solid state.

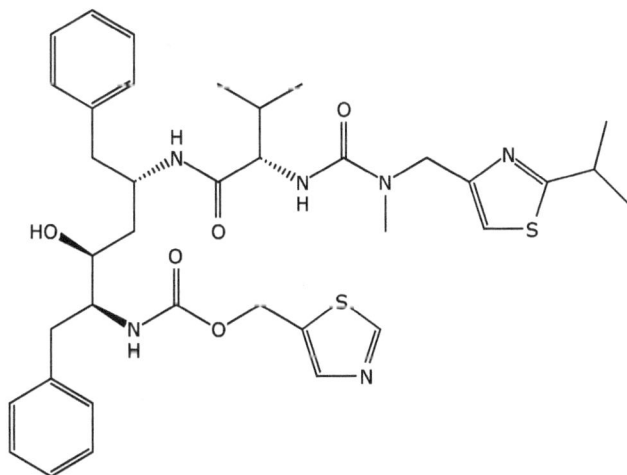

Fig. 4.5: Ritonavir: its manufacture was interrupted by formation of a new polymorph.

The polymorph nightmare first began in 1998, when capsules started failing the solubility test [183]. A new polymorph had formed, Form II, which was much less soluble than the original Form I. Shortly afterwards, the new polymorph was present throughout the formulation plant. Samples of Form II were taken into the lab for investigation, and it soon became impossible to produce Form I in that environment. Following a visit from staff exposed to Form II, the new polymorph started appearing in the bulk manufacturing plant [184]. It was no longer possible to produce the original formulation of ritonavir, and production had to be suspended despite the urgent requirement for the drug from AIDS patients.

Experimentally, it was actually quite difficult to produce Form II without a seed being present (either deliberately added or assumed to be present in the atmosphere from site contamination). It was, thus, not surprising that pre-production polymorph screens had only given Form I. This reluctance of Form II to form without a seed raises the question as to how Form II was produced in the first place. There is speculation that it may have been due to a product impurity acting as a template.

Both forms of ritonavir were characterised [183]. They had very similar melting points and could not be distinguished by DSC; however, IR, solid-state NMR and XRPD could all be used for this purpose. Single crystal X-ray scans allowed the two structures to be determined, and the different hydrogen bonds present could be deduced: greater hydrogen bonding in Form II seems to be the main factor in giving it increased stability.

New formulations of ritonavir were produced that were compatible with Form II, such as refrigerated gel caps and tablets. In addition, methods were developed [184] to produce either form, as required. Form I could be made by dissolving ritonavir in ethyl acetate, filtering the hot solution through a filter cartridge and adding it to a mixture of Form I crystals and heptane antisolvent. Alternatively, ritonavir could be dissolved in a mixture of ethyl acetate and heptane, prior to seeding with Form I. The latter process was carried out on a large scale to give Form I with less than 3% Form II being present. Careful control of solvent volumes and temperature was required.

4.4 Case study 2 – aspirin

Aspirin (acetylsalicylic acid, Fig. 4.6, **h**) was first marketed by Bayer, back in the 1890s, and has remained a best-selling drug ever since. For decades, only a single polymorph was known. Form II of aspirin was discovered in the 1960s [185], but it was only in 2005 that a full characterisation was reported [186]. It turned out that the structure and properties of Form II were similar to the initial Form I. Indeed, most samples of Form II were actually mixtures with Form I, an inter-grown phase containing both polymorphs typically being formed. It was eventually discovered that pure crystals of Form II could be obtained [187] by crystallisation of aspirin from tetrahydrofuran or acetonitrile in the presence of aspirin anhydride (**i**).

Another polymorph, Form III, could be reduced at high pressures, but reverted to Form I at ambient pressure [188]. Neither of these new forms was commercially significant.

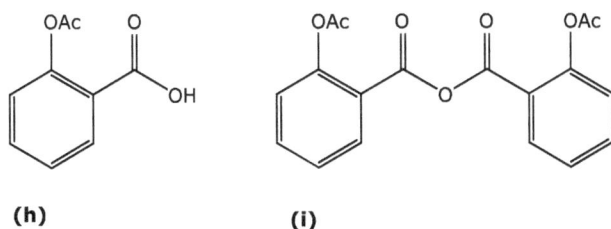

(h) **(i)**

Fig. 4.6: Aspirin (acetylsalicyclic acid, **h**) and aspirin anhydride (**i**).

Recently, a new form of aspirin was produced by melting crystals between glass slides by a team led by Chunhua (Tony) Hu from New York University [189]. The cooled melt gave spherulites of Form I and also spherulites of the new form (Form IV), the latter making up about 15% of the solid. Since the spherulites contained molecules organised in a helical structure, they could be distinguished by polarised light microscopy, despite aspirin being an achiral compound.

Form IV reverted to Form I after a few minutes at room temperature, but could be stabilised to last for around an hour by mixing with a suitable stabiliser, such as mannitol, Canadian balsam or polyvinylpyrrolidone. The new form was stable for around a day at 4 °C. Both Raman spectroscopy and XRPD could be used to distinguish it from Form I. Using specialist equipment and some complicated calculations, it was possible to use X-ray analysis to determine the structure, which was distinct from Forms I and II. Calculations showed that Form IV had a higher energy (by around 8 kJ/mol) than Forms I and II, with the latter two polymorphs having almost equal energy.

The authors state that the new form of aspirin should have greater bioavailability. Thus, it could be a medically useful if the stability and formulation challenges could be overcome. In general, metastable forms can offer the potential for more active formulations, provided reversion to the more stable polymorph can be avoided. An example of higher bioactivity is the recent discovery by the same team of a new polymorph of the insecticide deltamethrin [190], which is more active against mosquitos than the more stable form, and hence could be of great use in the control of malaria.

4.5 Case study 3 – atorvastatin calcium

Atorvastatin calcium (Lipitor, Fig. 4.7) is a true 'blockbuster' drug, one of the best-selling drugs of all time, and demonstrates a complex array of polymorphs and associated patents. It was initially developed by Warner-Lambert, which was later taken over by Pfizer. The initial patent for the calcium salt (US 5273995) [191] expired in 2011 in the USA, but a later patent (US 5969156) [192] on the main crystalline forms (Forms I, II and IV) was used by Pfizer in an attempt to prevent generic companies entering the market. Numerous court cases followed in various jurisdictions: atorvastatin also proved to be a blockbuster for the legal profession.

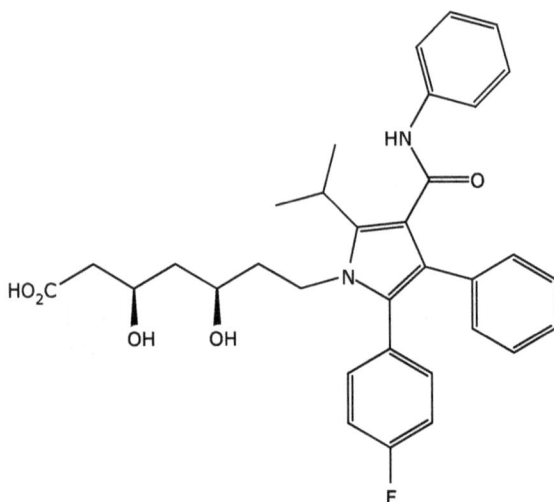

Fig. 4.7: Atorvastatin (typically used as the calcium salt).

Form I (a trihydrate) is the main commercial form, used by Warner-Lambert (Pfizer) and many generic companies. It seems to be stable in all its formulations. An examination of six commercial samples of crystalline atorvastatin calcium from different Indian manufacturers using XRPD and DSC showed that five were pure Form I, while one was Form I contaminated with other polymorphs [193]. Various other crystalline polymorphs, such as Teva's Form VIII have been used commercially, the latter having been quantified using Raman spectroscopy [194]. The amorphous solid has also been employed commercially [193], avoiding patents on the crystalline forms. A selection of the numerous patents and papers describing a multitude of polymorphs is displayed in Tab. 4.1.

On the whole, the forms from different companies in Tab. 4.1 differ, even when they share the same number (e.g. Form VIII from Warner-Lambert is not the same as Form VIII from Teva, the XRPD spectra being quite different). A vast array of methods was employed to make the various forms, using a wide range of solvents. In addition to the forms listed, various other solvates [218] and co-crystals have been claimed over the years, as well as salts other than the calcium salt. The Warner-Lambert patent [192] on Forms I, II and IV has now expired, so the situation is easier for generic companies wishing to produce atorvastatin calcium. Few, if

Tab. 4.1: Selected patents and papers on atorvastatin calcium polymorphs.

Patent/paper	Assignee or author affiliation	Form described	Date of publication
US 4681893 [195]	Warner-Lambert	Initial patent (for atorvastatin)	1987
US 5273995 [191]	Warner-Lambert	Initial patent for Ca salt (amorphous)	1993
US 5969156 [192]	Warner-Lambert	Forms I, II and IV	1999
EP 848704 [196]	Warner-Lambert	Form III	2001
WO 02/057229 [197]	Biocon	Form V	2002
WO 03/050085 [198]	Ivax	Forms Fa and Je	2003
US 6528660 [199]	Ranbaxy	Amorphous (Process patent)	2003
US 6605729 [200]	Warner-Lambert	Forms V to XIX	2003
US 6646133 [201]	Egis	Amorphous (Process patent)	2003
WO 2004/050618 [202]	Teva	Form F	2004
WO 2005/090301 [203]	Ranbaxy	Form R	2005
EP 1562583 [204]	Morepen Laboratories	Form VI	2005
US 6992194 [205]	Teva	Forms VI, VII, VIII, IX, X, XI and XII	2006
WO 2006/011041 [206]	Warner-Lambert	Forms XX to XXX	2006
WO 2006/012499 [207]	Teva	Forms XVIII and XIX	2006
WO 2006/048894 [208]	Morepen Laboratories	Forms M-2, M-3 and M-4	2006

Tab. 4.1 (continued)

Patent/paper	Assignee or author affiliation	Form described	Date of publication
WO 2006/106372 [209]	Egis	Form B-52	2006
US 7074818 [210]	Dr Reddy's	Forms VI and VII	2006
WO 2007/096903 [211]	Matrix Laboratories	Form M	2007
US 7411075 [212]	Teva	Form V	2008
An et al. [213]	Duksung Women's University	Forms 2 and 3	2008
US 7501450 [214]	Teva	Forms VI, VII, VIII, IX, IXa, X, XI, XII, XIV, XVI, XVII	2009
US 7538136 [215]	Teva	Forms X, A, B1, B2, C, D and E	2009
US 8080672 [216]	Teva	Form T1	2011
Rao et al. [217]	Aurobindo and Acharya Nagarjuna University	'New polymorphic form'	2011

any, compounds have quite such a complex array of polymorphs as atorvastatin, but many process development chemists will have the joy of sorting out a large smorgasbord of polymorphic forms at some point in their career.

4.6 Prediction of polymorph structures

With today's ever-increasing computational power, many researchers have attempted to predict polymorphic structures and their relative stability by theoretical means [219, 220]. This discipline is referred to as crystal structure prediction. A large number of possible crystal structures can be generated, and their lattice energies estimated and compared. In many cases, such as caffeine (Fig. 4.8, **j**), a number of forms with low lattice energy are given by calculation [221], and at least three anhydrous forms and one hydrate exist in practice [222]. In contrast, isocaffeine (**k**) has one form with lattice energy around 6 kJ/mol less than its nearest rival. This large energy difference is relatively unusual and is also in line with experiment, since only one polymorph of isocaffeine has ever been isolated.

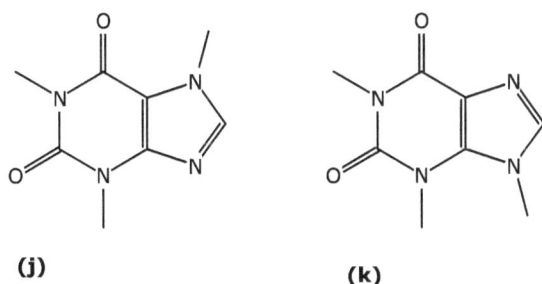

(j)   **(k)**

Fig. 4.8: Caffeine (j) and isocaffeine (k).

Various computational methods are used to calculate lattice energies. An example is density functional theory [223, 224], which has been applied to a number of organic compounds. Calculating lattice energies, although useful, gives results for 0 K, which may not always apply at ambient temperatures. Ideally, the relative free energies of theoretical polymorphs are needed as a function of temperature, but accurate estimation of such values is difficult. In recent years, molecular modelling techniques have been applied to temperature-dependent polymorphic changes [225], which usually occur because of differences in entropy. However, even if you know that a certain structure will be the most stable at a certain temperature, kinetic factors may mean that it is hard to isolate it in practice.

It is necessary to objectively compare the different computational methods used to find polymorphs. Every few years, the Cambridge Crystallographic Data Centre organises a series of blind tests for the various groups working in the field, to find out how each performs in polymorph prediction for a number of compounds. At the time of writing, the seventh blind test is running. The sixth blind test [226] involved five target compounds with unpublished crystal structures, varying in their flexibility, charges, etc. All of the experimental structures were predicted by at least one submission. On the other hand, no group successfully predicted all five structures. Over the years, the ability of programs to predict crystal structures has steadily improved. It is no longer sensible for the practical chemist to ignore theoretical inputs with regard to polymorphs, so it may be best to bring in an expert in the field as a consultant. Empirical testing can only take you so far: at some point you will need to decide whether it's worth continuing with attempts to form new polymorphs or not, and theoretical calculations are increasingly able to provide guidance on this question.

Chapter 5
Optimisation and experimental design

5.1 Why use experimental design?

Process optimisation is the central core of process development. In addition to improving product yield and purity, other so-called output variables (also referred to as dependent variables), such as particle size, polymorphic form or levels of residual metals may also need to be optimised. The conditions of the experiment, made up of the so-called input variables (also referred to as independent variables), such as temperature, solvent and proportion of reagent, are adjusted in order to achieve optimisation. The old-fashioned way of carrying out optimisation is by starting with a standard method (maybe from the research department or the literature) and changing one variable at a time. So a run is carried out at a higher temperature, keeping everything else constant; another run is then carried out with a higher stirrer speed (with the temperature returned to its original value); and a third run is carried out with a different reagent (with temperature and stirrer speed both at the original value), etc. This procedure is known as 'one factor at a time' (abbreviated as OFAT), and despite all attempts to kill it off, it survives like Frankenstein's monster in many laboratories to this day.

One disadvantage of OFAT is that it is common for input variables to interact with each other. Thus, in the example given, the new reagent may only be more effective than the original reagent at a higher temperature, while giving lower yields at the original temperature. Such interactions are not detected by the OFAT approach, with the experimenter completing their 'optimisation' in blissful ignorance of their existence [227]. In contrast to OFAT, experimental design (also known as 'design of experiments', DoE) aims to cover a volume of 'experimental space' with the optimum number of experimental runs, obtaining a comprehensive overview of the effects of different variables and their interactions, which can be expressed by an algebraic equation if necessary. Thus, the important input variables can be distinguished from those that make little difference, allowing for their further optimisation using a second experimental design if required. Understanding the 'experimental space' allows robust conditions to be employed, where small perturbations will not lead to a significant drop in yield or purity. A good experimental design will also allow a better estimation of the random errors involved in the runs, allowing accurate determination of the likelihood that a particular effect is statistically significant.

Experimental designs have proved their worth over a vast range of disciplines and are widely used in many industries. They are nowadays a mainstay of chemical process development. A recent example [228] of optimisation by experimental design from Gilead scientists involved a selective continuous DIBAL reduction of a

https://doi.org/10.1515/9783110717877-006

diester (Fig. 5.1 **a**) to the mono-aldehyde (**b**). Looking at the effects of temperature and equivalents of DIBAL, it was found that a relatively high temperature and low amount of DIBAL gave the best yields.

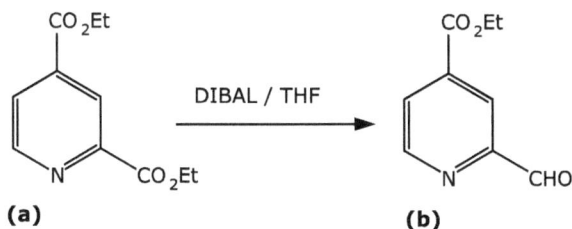

Fig. 5.1: Conversion of diester (**a**) to mono-aldehyde (**b**), optimised by experimental design.

Another recent example of experimental design was its use to optimise the drying of a cyclodextrin active pharmaceutical ingredient from Merck [229], examining the effect of humidity, temperature and pressure. Conditions were found that gave levels of solvent that met the specification. In a third example, Hovione chemists recently [230] used an experimental design to optimise a palladium-catalysed Sonogashira coupling reaction, examining the effect of various input variables. Thus, it can be seen that today's industry regularly uses experimental designs to tackle a wide range of problems.

5.2 Factorial and partial factorial designs

Factorial and partial factorial designs [231, 232] are powerful techniques for optimisation and are commonly applied to a wide range of process development problems. Considering the example mentioned previously of examining temperature (given as a 5 °C range, since it is hard to keep to an exact temperature), stirrer speed and different reagents, let's suppose that OFAT runs gave the following values for yield (Tab. 5.1).

Tab. 5.1: Typical OFAT runs.

	Temperature (°C)	Reagent (X or Y)	Stirrer (rpm)	%Yield
Run 1 (initial)	20–25	X	200	60
Run 2	60–65	X	200	55
Run 3	20–25	Y	200	52
Run 4	20–25	X	400	59

The various changes to the initial conditions have not led to any improvement. However, improvement is still possible if one or more interactions are significant.

By adding four additional runs, we obtain a balanced experimental design and the following yields (Tab. 5.2).

Tab. 5.2: An experimental design (2^3 factorial).

	Temperature, T (°C)	Reagent, R (X or Y)	Stirrer speed, S (rpm)	%Yield
Run 1	20–25	X	200	60
Run 2	60–65	X	200	55
Run 3	20–25	Y	200	52
Run 4	20–25	X	400	59
Run 5	20–25	Y	400	53
Run 6	60–65	Y	200	75
Run 7	60–65	X	400	54
Run 8	60–65	Y	400	75

Here, it can clearly be seen that reagent Y gives much better yields, but only at higher temperatures, while the stirrer speed does not produce a significant effect in this case. The above experimental design is designated as a 2^3 factorial design, since three input variables are examined at two different levels [231]. The eight runs in the 2^3 factorial design makes up the corners of a cube (Fig. 5.2).

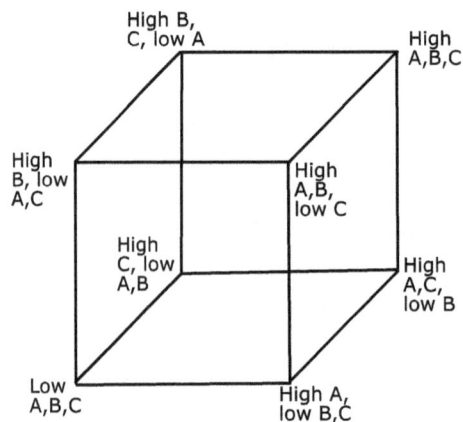

Fig. 5.2: Representation of 2^3 factorial.

A 2^3 factorial design, such as the one shown in Tab. 5.2, can be described by a matrix (sometimes referred to as an analysis matrix or design matrix), with the input variables represented by either −1 or +1, depending on their level, as shown in Tab. 5.3 (T, temperature; R, reagent; and S, stirrer speed). Of course, for non-numerical input variables, such as catalyst type, the designation as −1 or +1 is entirely arbitrary, but does not affect the overall results. The interactions TR, TS, RS and TRS can also be included, being the multiples of the values for the single variables. This type of full factorial design is known as an orthogonal design, i.e. the effect of any one factor balances out when considered across the range of any other factor, so neither the factors nor their multiples correlate with each other at all.

Tab. 5.3: The experimental design from Tab. 5.2 described as a matrix with +1 and −1.

	T	R	S	TR	TS	RS	TRS	%Yield
Run 1	−1	+1	−1	−1	+1	−1	+1	60
Run 2	+1	+1	−1	+1	−1	−1	−1	55
Run 3	−1	−1	−1	+1	+1	+1	−1	52
Run 4	−1	+1	+1	−1	−1	+1	−1	59
Run 5	−1	−1	+1	+1	−1	−1	+1	53
Run 6	+1	−1	−1	−1	−1	+1	+1	75
Run 7	+1	+1	+1	+1	+1	+1	+1	54
Run 8	+1	−1	+1	−1	+1	−1	−1	75

The net effect of T can be determined by adding or subtracting the yields, depending on whether the T value was +1 or −1, and then dividing by four (four 'differences' between yield values). A similar procedure can be applied to the other input variables and their multiples.

Effect of $T = (−60 + 55 − 52 − 59 − 53 + 75 + 54 + 75)/4 = +35/4 = +8.75$
Effect of $R = (+60 + 55 − 52 + 59 − 53 − 75 + 54 − 75)/4 = −27/4 = −6.75$
Effect of $S = (−60 − 55 − 52 + 59 + 53 − 75 + 54 + 75)/4 = −1/4 = −0.25$
Effect of $TR = (−60 + 55 + 52 − 59 + 53 − 75 + 54 − 75)/4 = −55/4 = −13.75$
Effect of $TS = (+60 − 55 + 52 − 59 − 53 − 75 + 54 + 75)/4 = 1/4 = −0.25$
Effect of $RS = (−60 − 55 + 52 + 59 − 53 + 75 + 54 − 75)/4 = −3/4 = −0.75$
Effect of $TRS = (+60 − 55 − 52 − 59 + 53 + 75 + 54 − 75)/4 = +1/4 = +0.25$

From these coefficients, the yield can be determined by the following equation:

$$\text{Yield} = 60.375 + (8.75T/2) − (6.75R/2) − (0.25S/2) − (13.75TR/2)$$
$$− (0.25TS/2) − (0.75RS/2) + (0.25TRS/2)$$

The values T, R, S, etc. are either -1 or $+1$, according to the matrix. Each of the various effects are divided by 2 since there is a difference of 2 between the -1 and $+1$ used to represent the levels of the input areas. The value of 60.375 seen in the equation is the overall mean yield. Such equations are referred to as models since they can be used to estimate the yield. From the above model, we can confirm that TR is the most crucial factor in determining the yield, while the stirrer speed is unimportant. Such equations are only estimates and are distorted by experimental error, which can be much reduced by running each experiment more than once. However, if you duplicate each run, you have to carry out 16 runs rather than eight. Error calculations will be discussed further later in this chapter.

Of course, there are many statistical programs that will do such calculations for you, which is especially useful when a larger number of factors need to be considered. These include 'Design-Expert' from Stat-Ease, JMP's 'Design of Experiments', Minitab software and Statgraphics software. Randomisation of the runs is important in experimental design. You don't want any of your main effects, such as temperature, stirrer speed and reagent in our example, to first be at one value for four runs and then at the other for the next four runs. Programs will randomise the order in which you carry out the reactions, although it is worthwhile checking that the order is not highly correlated with any of the main input variables (since randomisation does not always prevent unwanted correlations) and correcting if necessary. You may get more proficient (or more careless) at running the reaction as the experimental design proceeds. Also, there may be a need to replace a batch of reagent, etc. if the material runs out before the experiment is completed, which again may affect the overall outcome.

The 2^3 factorial design described above is fine when you are sure that only three input variables are important. However, an initial design often has to consider more input variables. It may well be that some of these turn out to be unimportant and a second design can be run that leaves them out, but it's best not to leap to conclusions before looking at the problem in more detail. In practice, it is likely that four to ten input variables will have to be initially considered for a typical reaction. A 2^4 factorial design requires 16 runs, while a 2^{10} requires 1,024 runs. Fortunately, it is still possible to obtain useful results using fewer runs. In the above example, it is clear that TR represents the interaction of T and R, but what exactly does TRS mean in real terms? With more runs, much of the acquired information from a full factorial run gives coefficients for such multi-factor variables, which, in practice, rarely have any importance, since the output is typically determined by the main effects (e.g. T and R in our example) and two-factor interactions (e.g. TR in our example). Three-factor (or higher factor) variables, such as TRS, can be 'aliased' with the main input parameters, meaning that their effects cannot be distinguished. These so-called fractional factorial or partial factorial designs allow far fewer runs to be used than a full factorial design. In addition, some two-factor interactions can be aliased if common sense tells you they are unlikely. Consider a strongly basic reaction where, in addition to the reaction parameters, you are interested in whether the completed reaction should be quenched with ammonium chloride solution (research department method) or acetic acid solution (proposed new method to avoid chloride ions). You might decide that interactions with the reaction temperature or reaction stirrer speed would be unlikely, although an input parameter such as choice of solvent might possibly interact. Partial factorials can be determined mathematically, but a good design of experiments program will do all the hard work for you, giving suitable designs for a given number of runs.

5.3 Analysis of variance (ANOVA)

The statistical analysis of factorial and partial factorial results is usually carried out using the so-called analysis of variance or ANOVA [231, 232]. This procedure compares the variation due to the different treatment levels to each other and to the background random variation (referred to as 'residual variation' or 'variation due to error'). The effect of input variables, including those multi-factor variables that are set up to be examined (AB, AC, etc.), is calculated to give a p value, which is a measure of significance. Typically, variables with a p value of <0.05 are considered significant. This p value is referred to as the 'level of significance', α, so when using statistical programs, it may be necessary to set α to 0.05. Formally, this can be stated as being 95% certain that we can reject the so-called null hypothesis, which states that the particular variable has no effect on the output variable within the range examined. Of course, it would be unwise to ignore a result of say $p = 0.06$, which obviously gives some indication that a particular level of a variable may be favoured. Typically, for $0.05 < p < 0.10$, we can conclude that the variable may be better at the indicated level and proceed on that assumption. Process development needs to be a carried out speedily, so it is often not possible to spend a long time obtaining a statistically perfect result. However, if there are good practical reasons for preferring the level that is not indicated by the results, such as a lower cost, you may just have to carry out further runs, reducing the overall random error so that the results become clearer. If there is true significance, p should eventually shift to <0.05 if sufficient runs are carried out.

As an example of an ANOVA analysis, we can consider a hypothetical 2^2 factorial experiment, looking at the effects of two input factors, the excess of reagent, A (either 1.1 or 1.5 equivalents), and addition time, B (either 1 or 3 h), on yield. In practice, a typical 'first look' factorial experiment would usually contain more than just two factors, but having a simple example makes the calculations clearer. Two replicate runs were carried out for each of the four sets of conditions, giving the results shown in Tab. 5.4.

In Tab. 5.4, runs 1 and 2, 3 and 4, etc. are replicates, allowing random variation to be estimated. The above results clearly show a strong effect for variable A, with 1.5 equivalents of reagent being favoured. However, chemists may argue whether there is a slight effect for B or not, so further examination is required. Table 5.5 gives the analysis matrix for these results.

Tab. 5.4: 2^2 factorial design, with two replicates (runs not randomised, in order to improve clarity).

	A (equivalents)	B (hours)	%Yields
Run 1	1.1	1	70
Run 2	1.1	1	72
Run 3	1.5	1	81
Run 4	1.5	1	82
Run 5	1.1	3	73
Run 6	1.1	3	71
Run 7	1.5	3	82
Run 8	1.5	3	86

Tab. 5.5: Analysis matrix for 2^2 factorial design with two replicates.

	A (equivalents)	B (hours)	AB (interaction)	%Yields
Run 1	−1	−1	+1	70
Run 2	−1	−1	+1	72
Run 3	+1	−1	−1	81
Run 4	+1	−1	−1	82
Run 5	−1	+1	−1	73
Run 6	−1	+1	−1	71
Run 7	+1	+1	+1	82
Run 8	+1	+1	+1	86

The effects of the two factors and their interaction can be calculated as follows:

Effect of $A = (81 + 82 + 82 + 86 − 70 − 72 − 73 − 71)/4 = 45/4 = 11.25$
Effect of $B = (73 + 71 + 82 + 86 − 70 − 72 − 81 − 82)/4 = 7/4 = 1.75$
Effect of $AB = (70 + 72 + 82 + 86 − 81 − 82 − 73 − 71)/4 = 3/4 = 0.75$

The so-called 'sum of squares' (SS) provides a useful measure of the amount of the variation due to the various effects and the random error. The sum of squares for the effects in a 2×2 factorial experiment can be calculated using the formula $SS_A =$ (sum of results for $A)^2/4n$ (where n is the number of replicates, in this case 2).

Thus, the following SS values can be found for the effects:

$SS_A = 45^2/8 = 253.125$
$SS_B = 7^2/8 = 6.125$
$SS_{AB} = 3^2/8 = 1.125$

However, in addition to the sum of squares for effects, we can also calculate a sum of squares for the error (SS_E, also known as the 'residual sum of squares', SS_{RES}). For each run, the difference from the mean for the two replicates is squared, giving:

Run 1: Value = 70, mean for replicates = 71, difference squared = $1 \times 1 = 1$
Run 2: Value = 72, mean for replicates = 71, difference squared = $1 \times 1 = 1$
Run 3: Value = 81, mean for replicates = 81.5, difference squared = $0.5 \times 0.5 = 0.25$
Run 4: Value = 82, mean for replicates = 81.5, difference squared = $0.5 \times 0.5 = 0.25$
Run 5: Value = 73, mean for replicates = 72, difference squared = $1 \times 1 = 1$
Run 6: Value = 71, mean for replicates = 72, difference squared = $1 \times 1 = 1$
Run 7: Value = 82, mean for replicates = 84, difference squared = $2 \times 2 = 4$
Run 8: Value = 86, mean for replicates = 84, difference squared = $2 \times 2 = 4$

The SS_E value is given by the sum of the results for each run, $1 + 1 + 0.25 + 0.25 + 1 + 1 + 4 + 4 = 12.5$.

An overall sum of squares, SS_T, can also be calculated, using the overall mean (also known as the 'grand mean') from all eight runs, μ, which was equal to 77.125 for this example. The overall sum of squares is given by the sum of the square of the differences between the overall mean and the value of each run, giving:

Run 1: Value = 70, difference from μ = 7.125, difference squared = 50.76563.
Run 2: Value = 72, difference from μ = 5.125, difference squared = 26.26563
Run 3: Value = 81, difference from μ = 3.875, difference squared = 15.01563
Run 4: Value = 82, difference from μ = 4.875, difference squared = 23.76563
Run 5: Value = 73, difference from μ = 4.125, difference squared = 17.01563
Run 6: Value = 71, difference from μ = 6.125, difference squared = 37.51563
Run 7: Value = 82, difference from μ = 4.875, difference squared = 23.76563
Run 8: Value = 86, difference from μ = 8.875, difference squared = 78.76563

Thus, SS_T = the sum of the above 'difference squared' values = 272.875 (to 3 decimal places)

The SS values for the effects plus that for the error give SS_T, i.e. $SS_A + SS_B + SS_{AB} + SS_E = SS_T$ (253.125 + 6.125 + 1.1.25 + 12.5 = 272.875). Thus, the effect of A accounts for almost 93% of the total sum of squares. The relative low value of SS_E (error SS or residual SS) shows a well-run experiment, with good control over random variation.

We can now calculate the mean square, MS, for the effects and error. This quantity is simply the SS divided by the relevant degrees of freedom (D of F), which will be 1 for A, B and AB (since there are only two levels; with three levels, there would be two degrees of freedom, etc.). For the error, the degrees of freedom = 4. The degrees of freedom for the error of a 2^2 factorial design are given by the formula D of F = 4(n − 1), where n = the number of replicates, while for a 2^3 design the formula is D of F = 8 (n − 1).

From the MS statistics, it is possible to calculate the so-called F value for each effect, including their combinations, by simply dividing the MS value by the MS for the error. F is used to determine whether or not an effect is significant at a particular level of significance (usually α = 0.05). If the calculated F value for an effect is more than the critical F value (either computed or obtained from tables [231]), we can say that the effect is significant (more formally, we can reject the null hypothesis that there is no difference between the levels of that particular effect). ANOVA results tables typically also include p values for the effects. Thus, a typical ANOVA results table for the above example would look something like Tab. 5.6.

Tab. 5.6: ANOVA results table for 2^2 factorial experiment.

Source of variation	SS	Degrees of freedom	MS	F value	F (critical)	P
A	253.125	1	253.125	81.00	7.71	0.0008
B	6.125	1	6.125	1.96	7.71	0.23
AB	1.125	1	1.125	0.36	7.71	0.58
Error	12.5	4	3.125			
Total	272.875	7				

The exact appearance of an ANOVA results table will vary, depending on the computer program used. Of course, few people will carry out an ANOVA calculation one step at a time as above, since many suitable programs are available, but doing so illustrates where the numbers come from, which helps in understanding the output. The results in Tab. 5.6 confirm the highly significant effect of A, while the interaction AB is clearly unimportant in this case. The effect of B is not even significant at α = 0.10, so we have no grounds for rejecting the null hypothesis that addition time (B) makes no difference. That is not to say that if enough replicates were carried out, a significant relationship might not be found. However, it is likely that resources would be better employed elsewhere, since any effect of B is clearly small. In general, small effects that are not statistically significant can usually be ignored. On the other hand, if an effect is large, but not quite significant, further runs may be needed to decide whether there is true significance or not.

Adding the grand mean to the coefficients for the effects calculated earlier gives us the following model equation for the experiment:

$$\text{Yield} = 77.125 + (11.25A/2) + (1.75B/2) + (0.75AB/2)$$

The differences between the yield values given by the model and the actual yield values are the residuals, the values of which arise due to random error. These should roughly follow a normal distribution and not correlate with the order in which the runs were carried out. If either of these statements does not apply, you may well have a problem. Thus, for our experiment, Tab. 5.7 shows the differences between the actual values and those derived from the model equation (in this case, these calculated values are equal to the mean for each pair of replicates).

The order of runs shown in Tab. 5.4 and Tab. 5.5 was given for reasons of clarity, but in practice the runs would not have been carried out in this order since it correlates completely with factor B. Let us suppose that the order of runs was in fact run 7, run 1, run 4, run 8, run 2, run 5, run 6 and run 3, which has little correlation with the main effects.

Tab. 5.7: Table of residuals.

Run (actual order in brackets)	Yield (actual)	Yield (calculated)	Residual
Run 1 (2nd)	70	71	−1
Run 2 (5th)	72	71	1
Run 3 (8th)	81	81.5	−0.5
Run 4 (3rd)	82	81.5	0.5
Run 5 (6th)	73	72	1
Run 6 (7th)	71	72	−1
Run 7 (1st)	82	84	−2
Run 8 (4th)	86	84	+2

The Pearson's correlation coefficient, R, between the actual run order and the residuals is 0.24, showing no cause for concern. A high coefficient (i.e. one near to −1 or +1) would suggest changes in the reaction with time. This could be due to many reasons, a common one being using up an old batch of starting material or reagent and replacing it with a new one. In this case, purity or hydration levels may differ, which may affect the reaction.

A histogram of the frequency of occurrence of residual values against these values (rounded, if necessary) should give a roughly bell-shaped normal distribution. A histogram of the residuals is shown in Fig. 5.3, with the frequency given for each range (ranges −2 to <−1, −1 to <0, 0 to +1, and >+1 to +2), giving a roughly normal distribution. An experiment with a greater number of runs should give a distribution closer to normality. So-called normal probability plots can also be used to analyse residuals [232].

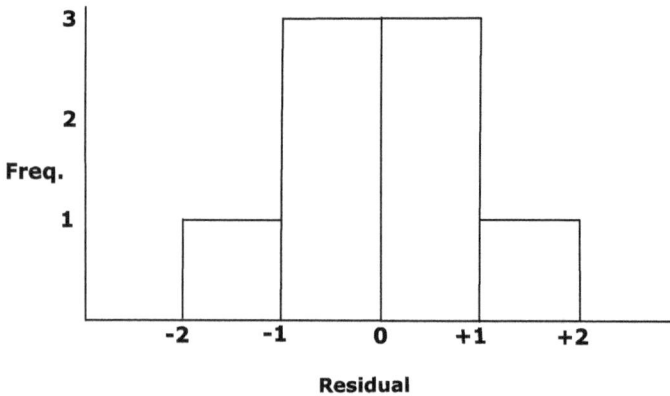

Fig. 5.3: Residual values from Tab. 5.7 plotted as a histogram.

We can now consider the experiment described previously, with three input variables, T, R and S, but with no replicates. The results were described for the three variables and four multiples (TR, TS, RS and TRS), making seven effects in all. However, since there are a total of seven degrees of freedom and each effect uses one degree of freedom, there are no degrees of freedom left for the error. The yield is given by the model equation, seemingly without any way to determine the error. So, it appears that we can say nothing about the accuracy of our results, which feels intuitively wrong.

One possible solution, as mentioned previously, is to carry out replicate runs, but it may be the case that there is a shortage of starting material or no time to carry out further experimentation. It is possible to ignore the effect of the triple multiple TRS since its magnitude is small and such triple multiples are rarely significant. This would leave one degree of freedom for the error. Another solution, although frowned upon by some statistical purists, would be to totally ignore the effect of S and its multiples since they are all small and changing the 2^3 design with no replicate runs to a 2^2 design with two replicates runs for each combination.

A better method of dealing with the problem would be to add the variables, one at a time, to an equation, starting with the most significant. The three most significant variables were TR, T and R, with coefficients of -13.75, $+8.75$ and -6.75, respectively.

The original complete equation for the yield was given by

$$\text{Yield} = 60.375 + (8.75T/2) - (6.75R/2) - (0.25S/2) - (13.75TR/2)$$
$$- (0.25TS/2) - (0.75RS/2) + (0.25TRS/2)$$

Taking only T, R and TR coefficients gives:

$$\text{Yield} = 60.375 + (8.75T/2) - (6.75R/2) - (13.75TR/2)$$

We could then perhaps add the next highest coefficient, 0.75 for RS, to give:

$$\text{Yield} = 60.375 + (8.75T/2) - (6.75R/2) - (13.75TR/2) - (0.75RS/2)$$

The yields from these equations with fewer variables can be compared to the actual yield, which is given by the complete equation. The results are shown in Tab. 5.8:

Tab. 5.8: Adding variables to a model equation for T, R and S (see Tab. 5.3).

	T (temp)	R (reagent)	S (stirrer)	Actual %yield	%Yield from equation with just T, R and TR	%Yield from equation with just T, R, TR and RS
Run 1	−1	+1	−1	60	59.5	59.875
Run 2	+1	+1	−1	55	54.5	54.875
Run 3	−1	−1	−1	52	52.5	52.125
Run 4	−1	+1	+1	59	59.5	59.125
Run 5	−1	−1	+1	53	52.5	52.875
Run 6	+1	−1	−1	75	75.0	74.625
Run 7	+1	+1	+1	54	54.5	54.125
Run 8	+1	−1	+1	75	75.0	75.375

The differences (residuals) between the actual yields and those calculated from just T, R and TR are small (six residuals of 0.5 and two of 0). Using the values for RS gave only slightly smaller residuals (six residuals of 0.125 and two of 0.375). Thus, RS and the unused coefficients, S, TS and TRS, can be omitted, giving four degrees of freedom for error calculations. Many statistical programs allow coefficients to be added one by one, showing the decrease in residuals as coefficients are added and producing ANOVA tables for the results. A judgement needs to be made as to when it is no longer sensible to add further variables, there being little change in residuals, although chemical considerations have still to be borne in mind.

5.4 Experimental designs with three or more levels

The various two-level experimental designs discussed so far are extremely useful, but do not give information about a non-linear response. Of course, it is common for yield to increase with a variable, such as temperature, up to a certain point and eventually decrease again. In such cases, yield may be described by a quadratic equation, involving terms such as A^2 and B^2, as well as A and B. A factorial experiment, such as a 3^2 factorial, involving two factors each at three levels, can be used to look for quadratic responses. Such factorials can lead to a large number of runs. A 3^3 factorial with two replicates at each point would require 54 runs ($3^3 \times 2$). A number of other designs are often used, the so-called central composite design (CCD) [231, 233] being popular. This design involves a basic two-level factorial design along with a centre point and so-called 'star points' forming a cross or crosses from the centre point, either on the edges of the factorial shape or further out (often the star points are at the same distance from the centre point as the corners of the factorial points). Several replicates are run at the central point but none elsewhere. A basic CCD for two variables is shown in Fig. 5.4.

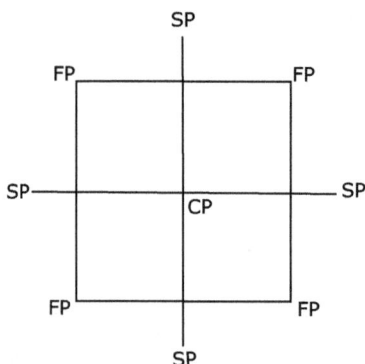

Fig. 5.4: CCD for two variables. Centre point, CP; factorial points, FP; star points, SP.

The central point is replicated four or five times, allowing the random error to be estimated. Such designs are useful for fine tuning the important variables after an initial two-level factorial or partial factorial design. For example, an initial design might tell you that yields were significantly higher at 60–65 °C than at 20–25 °C. You might then want to compare 60–65 °C to 50–55 °C and 70–75 °C, giving you three levels.

Experiments run with three or more levels allow the construction of contour plots, showing the levels of yield in a manner similar to contours on a map. A hypothetical example is given in Fig. 5.5. Superimposing contour plots for yield and purity can show regions where high levels can be found for both, which is particularly useful in reactions, such as some oxidations, where there tends to be a trade-off between these two output variables. It is important that the final process is on a 'plateau' and not on a 'cliff edge', as in the latter case small perturbations can lead to a catastrophic loss in yield or purity. Process development should aim at developing robust processes that will not be drastically affected by small changes in conditions.

Point P on the contour plot in Fig. 5.5 gives roughly the same yield as point Q, but the former is much to be preferred, since it is in the middle of a 'plateau', while Q is perched on a 'cliff edge'. In addition to contour maps, various three-dimensional response surfaces can be produced where more than two input variables have to be mapped.

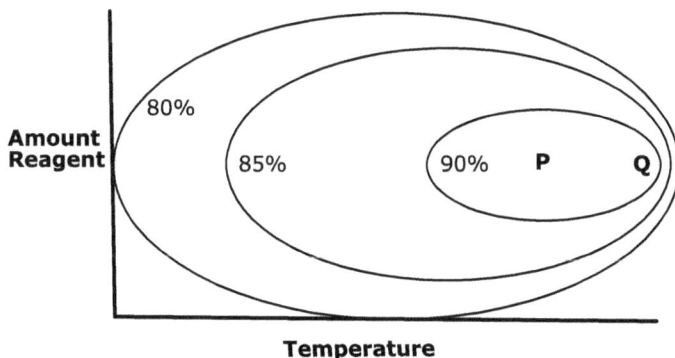

Fig. 5.5: Contour map, showing the effect on yield of the amount of reagent and temperature.

5.5 Choice of variables and reaction scale

There are a large number of possible experimental designs, and it is possible to spend a vast amount of time analysing your data by the myriad of available methods. However, the initial choice of variables often has far more effect on the outcome of an experimental design than the exact statistical methods used.

It is impossible to examine every possible variable by means of experimental designs, since process development needs to be carried out reasonably quickly. Imagine a reaction with one starting material, two reagents and two solvents. Even if we do not want to change these, just looking at samples from two different suppliers for each of these will give us five variables, and we have not even started the reaction yet. A good experience of plant chemistry is needed to select the variables that are likely to be important. In the example given, it would probably be best to just go with the commercial bulk samples (not samples from laboratory suppliers) that you plan to use, except where there is likely to be some variation. For example, potassium carbonate is a very useful base but notorious for giving somewhat different outcomes, depending on the exact grade used. It is probably worth obtaining samples of two different grades to put into your initial experimental design.

For a first factorial or partial factorial experiment, fairly large jumps in input variable values are more likely to show an effect, and a design with only small changes may well lead to inconclusive results. However, initial range-finding runs are often required. Consider a variable such as temperature. You might have a method that is carried out at 20 25 °C, but is rather slow. You could use temperature values of 20–25 °C and 60–65 °C in your design. It is a good idea to carry out an initial run at the higher temperature range, just to make sure you do not waste half your experimental runs making a tarry mess. Such range-finding experiments can be incorporated into the design if reasonable results are obtained.

Devising an initial experimental design requires you to carefully consider not only which factors to put in, but also what to leave out. The commonly significant factors include temperature, amount and type of reagent(s), addition time, stirrer speed, solvent and concentration, but sometimes one or more of these can be left out. For example, a slow reaction involving only miscible liquids is unlikely to be affected by stirrer speed, while a reaction involving insoluble solids may well be.

When the amount of starting material or regent is limited, it is important not to use it all up in one gigantic experimental design. The results may suggest that your initial runs are in the wrong area entirely. Consider a reaction that is carried out in the presence of a base. The literature suggests that excess base is required and the research department method uses 1.3 equivalents of base. You carry out a large design in which one factor is the level of base (either 1.0 or 1.3 equivalents). There is a clear advantage in using 1.0 equivalents. You use up your remaining gram of starting material on a run on Friday afternoon with 0.7 equivalents of base, and the yield surprisingly improves still further. Obviously, your initial design was in the wrong 'area', so you need to carry out another with lower base levels but you have no material left to do so. If the starting material is in short supply, it is a good rule of thumb not to use more than half of it in your initial experimental design.

Ideally, all experimental design should be carried out on the plant reactor to be used for the reaction, but this is seldom possible in practice, and is generally incompatible with the need for fast process development, even when starting material is abundant. Generally, a plant reactor is mimicked in the laboratory, using jacketed flasks with overhead stirrers and controlled feeds of liquids. Automated reactor control is invaluable to ensure accuracy and is discussed in more detail in the following chapter. However, if the supply of starting material is limited, you may end up using small round-bottomed flasks with magnetic stirrers. Random errors in yields tend to become greater as the reaction size decreases. For example, if you are weighing out 100 mg of solid, electrostatic factors can be a problem. It is easy to have a few mg of fine crystals stick to the inside of the flask neck without noticing. Making up a stock solution and using a micropipette may reduce errors somewhat, but even so, such very small-scale runs often simply produce random noise, all but the largest effects on yield being lost. It may be objected that successful small-scale laboratory designs are sometimes reported in literature, but the large number of unsuccessful ones are hardly likely to be mentioned. Nobody is going to tell the story of how they wasted a lot of time on a design that gave no clear answers.

Having said this, multiple small-scale reactions are very useful for the initial screening of catalysts, etc. Thus, a large number of palladium complexes and bases might be examined for a Suzuki–Miyaura coupling reaction. This is the domain of high-throughput experimentation (HTE), where multiple reactions are run simultaneously [234]. The goal is to find candidates for further investigation rather than mimic the yields from a plant process. Multiple catalysts or other reagents may be examined and (for example) the two most promising used for further development. Such approaches are further discussed in the following chapter.

Factors involved in the work-up, such as which washes to use, the amount of antisolvent, etc. need to be optimised. Such designs are best carried out as separate designs to the optimisation of the actual reaction. It's best to first optimise the reaction without work-up (apart from any necessary quenches), calculate the yield by HPLC or GC against a purified standard, and then carry out a large-scale reaction under optimum conditions to provide a stock solution or crude solid for the optimisation of the work-up. If it is problematic, drying conditions can be optimised separately, using a large batch of solvent-wet solid as input (this has to be carefully sealed to prevent solvent loss on standing).

5.6 Plant process improvements and CUSUMS

The job of a process development chemist does not necessarily stop once a process is successfully running on a manufacturing plant. Larger companies may have dedicated plant chemists to further improve the process, but in many cases such improvements are left, at least in part, to process development staff. Small increases in yield on a full-scale plant can make a huge difference to the profitability of a process, so should be sought where possible, while also bearing in mind any regulatory constraints.

In addition to increasing the yield, control of random deviation is essential for manufacturing processes. Even if the average yield and purity are high, too great a variability in yield or purity can be disastrous. Statistically, the standard deviation provides a useful measure of variability, and plant process improvement should aim for low standard deviation values, while any rise should be urgently investigated, before batches start failing to meet their specification.

Introducing a change to a plant process might seem straightforward. You compare the mean values of yield and purity for a number of runs before the process change is introduced, followed by a number of runs after. However, this view is naïve, since if no real change is made, you might still see a short-term improvement. The mere presence of extra professional staff on shifts, monitoring every detail of the process and interacting with the plant staff, could increase the yield or purity. Plant staff may be glad that somebody is taking an interest, so everything may be done exactly according to the book, tea breaks may become shorter, and tasks may be completed speedily rather than being left to the next shift. Such changes are sometimes referred to as the 'observer effect' or 'Hawthorne effect', named after the Hawthorne works where the effect was supposed to have occurred back in the 1920s, although subsequent reanalysis of the data suggests that possibly no such effect was actually present during these early studies [235].

In addition to a possible 'Hawthorne effect', the longer the time period over which a plant process is examined, the more likely that other changes may be introduced. For example, the original supplier may stop manufacturing a raw material, so a new supplier may have to be found. Changes to shift patterns may mean that a plant reaction mixture is left to stand for a longer period of time. Necessary repairs may lead to changes to vessels or other equipment. So you can end up with lots of plant data and more than one change. Such situations can be dealt with using 'cusums' [232, 236].

A cusum value for a particular batch is calculated by adding the deviation (which may be a positive or negative number) from the overall mean to the values from the previous batches, starting and ending at zero. This is best seen by examining the hypothetical example shown in Tab. 5.9. The output variable is yield in this example, but it could be product purity or impurity level. Let us suppose that our proposed improvement (increase in catalyst level) occurs prior to batch 10, while a change to the source of a raw material, due to the normal supplier ceasing to make it, occurred at batch 22. The overall mean was 90.5%.

Tab. 5.9: Cusums for plant batches with an overall mean yield of 90.5%.

Batch no.	Yield (%)	Deviation from mean	Cusum	Batch no.	Yield (%)	Deviation from mean	Cusum
1	90.3	−0.2	−0.2	21	90.1	−0.4	−8.6
2	89.6	−0.9	−1.1	22	90.6	0.1	−8.5
3	90.3	−0.2	−1.3	23	90.5	0.0	−8.5
4	90.0	−0.5	−1.8	24	91.0	0.5	−8.0
5	89.7	−0.8	−2.6	25	90.9	0.4	−7.6
6	90.2	−0.3	−2.9	26	91.1	0.6	−7.0
7	90.5	0.0	−2.9	27	90.8	0.3	−6.7
8	90.1	−0.4	−3.3	28	90.9	0.4	−6.3
9	89.6	−0.9	−4.2	29	89.9	−0.6	−6.9
10	90.2	−0.3	−4.5	30	91.3	0.8	−6.1
11	90.6	0.1	−4.4	31	91.4	0.9	−5.2
12	90.2	−0.3	−4.7	32	90.9	0.4	−4.8
13	90.3	−0.2	−4.9	33	91.2	0.7	−4.1
14	90.1	−0.4	−5.3	34	90.8	0.3	−3.8
15	89.5	−1.0	−6.3	35	91.0	0.5	−3.3
16	89.9	−0.6	−6.9	36	91.4	0.9	−2.4
17	90.5	0.0	−6.9	37	91.0	0.5	−1.9
18	89.6	−0.9	−7.8	38	91.3	0.8	−1.1

Tab. 5.9 (continued)

Batch no.	Yield (%)	Deviation from mean	Cusum	Batch no.	Yield (%)	Deviation from mean	Cusum
19	90.1	−0.4	−8.2	39	91.2	0.7	−0.4
20	90.5	0.0	−8.2	40	90.9	0.4	0.0

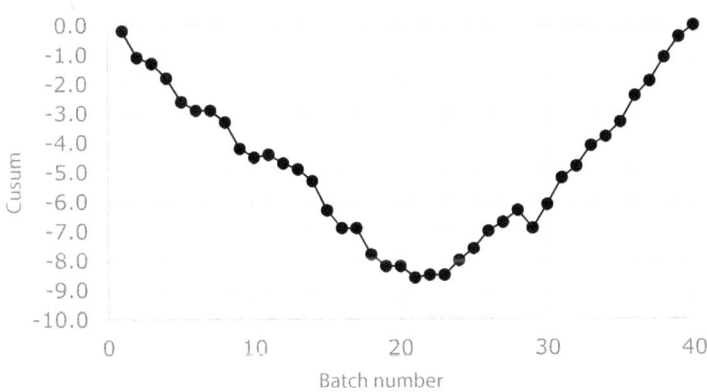

Fig. 5.6: Cusum graph, with process changes at batch 10 and batch 22.

A plot of the cusum values against batch number is shown in Fig. 5.6.

Changes in the yield are shown by changes to the slope of the cusum graph. The slope continues downwards up to around batch 22, after which it start going upwards. There is no great change to the slope due to our change in the amount of catalyst at batch 10, showing that, contrary to expectations, it made little difference. However, the forced change in raw material supplier at batch 22 clearly made a difference, giving a small increase in yield. The mean yield for batches 1–21 was 90.1%, while the mean yield for batches 22–40 was 91.0%. If we had simply compared the mean yield for the three batches (7–9) before the increase in catalyst level at batch 10 to the three batches after (10–12), we would have seen an increase in mean yield from 90.1% to 90.3%, but this was only due to short-term fluctuations, the cusum showing no prolonged change to the slope.

We can also check that there was a real change in yield at batch 22 from looking at the standard deviations. The overall standard deviation for the yield was 0.552, much higher than that for batches 1–21 (0.332) and batches 22 40 (0.356). Thus, the batches fall into two groups, separated by the change of conditions at batch 22. Overall, cusums can be a powerful method for examining large numbers of plant batches, highlighting where significant changes occurred.

Chapter 6
Automation

6.1 Automation for batch or continuous reactions

Nowadays, automated laboratory and pilot plant reactors are commonly used in process development. Automated control of temperature, pH, redox potential, input feeds and other parameters frees up a vast amount of staff time, while allowing greater reproducibility compared to manual control. Such automated systems can also allow the collection of valuable calorimetric data (see Chapter 2). Automation is particularly useful for experimental designs [237], typically reducing the random error between runs by allowing accurate control of the reaction conditions. Automated addition of liquids can be carried out using either syringe pumps or peristaltic pumps, the latter being particularly useful for larger scale reactions. Researchers from Glaxo and automation specialists Unchained Labs estimated that the use of automation decreased the working hours taken to optimise a particular reaction by four to five times [238].

Single- and multi-flask set-ups have been produced by a number of suppliers. The Atlas HD system from Syrris is an example of a single vessel system with a modular design, using jacketed flasks from 50 mL to 5 L and employing Syrris's software to allow reactions, crystallisations and calorimetric studies to be carried out under tightly controlled conditions. Another useful single reactor system is the H.E.L. AutoLAB II, which uses jacketed flasks from 250 mL to 50 L. H.E.L.'s WinISO software is employed for reactor control.

Mettler-Toledo's EasyMax systems are available as one-flask or two-flask systems (100 mL or 400 mL volume), while the related OptiMax system allows a volume of up to 1 L to be used. The Mettler-Toledo vessels integrate with many other instruments offered by this company, simplifying data handling.

For multiple vessels, Radleys' Mya4 system offers up to four reactors that are controlled independently, with flask sizes from 2 mL up to 400 mL. Their Mya control software allows for the control of both the reactors and third-party accessories, such as pH meters and syringe pumps. If budgetary constraints require you to produce your own automated system, a cheaper alternative is Radleys' Carousel 6 Plus, which allows six round-bottom flasks (5–250 mL) to be stirred with one stirrer hotplate. Here, the flasks are kept at the same temperature, rather than being independently controlled. Similar equipment for low temperature reactions (ice or cardice bath) is also available. Such inexpensive options are particularly useful in automated experimental designs, provided there is someone who can produce the necessary control system.

The H.E.L. AutoMATE II system is an advanced system allowing for up to eight flasks to be independently controlled, using either magnetic or overhead stirring. H.E.L.'s WinISO software is again used. This set-up is very useful for experimental

https://doi.org/10.1515/9783110717877-007

design and can also carry out multiple calorimetry experiments simultaneously. For the optimisation of catalytic reactions, H.E.L's ChemScan II allows up to eight simultaneous reactions to be carried out under pressure. Finding high-yielding catalysts often used to take weeks, but such equipment greatly speeds up the process. It is advantageous to take the two or three of the best-performing catalysts forward into a large-scale design, rather than just one, since their behaviour may differ somewhat on scale-up.

Continuous flow reactions [239], once mainly confined to bulk petrochemicals, are now routinely used in the fine chemical industry and can benefit greatly from suitable automation. Their safety advantages have been outlined in Chapter 2. Effective on-line analysis and control can help maintain output within the required parameters, allowing large quantities of product to be produced using relatively small reactors (process intensification). Continuous flow is particularly useful for photochemical reactions, the illumination being much more effective than shining a light into a large batch reactor, where most of the reaction volume is unaffected. A number of multistep automated flow reactions have been critically reviewed [240].

Vapourtec's R-series is a modular flow system, the components of which can be quickly changed (e.g. replacing a tubular reactor with a column reactor). It is controlled by R-series software, which can also be used to control other devices. One possible component of the R-series system is the UV-150 photochemical reactor, which can use either monochromatic LEDs or mercury lamps. The R-series also allows for flow reactions under pressure, using solvents above their normal boiling point. An example is the laboratory synthesis of 2,2-dimethylchromans (Fig. 6.1, **a**) from phenols (**b**) and isoprene in dichloromethane using a tropylium catalyst at 100 °C and 16 bar [241] by Thanh Vinh Nguyen's group at the University of New South Wales. Of course, a halogenated solvent is not recommended for industrial production, but the reaction shows the benefits of using solvents under pressure. Not surprisingly, the equivalent batch reaction gave too much polymerisation of isoprene, but high yields were achieved under the flow conditions used.

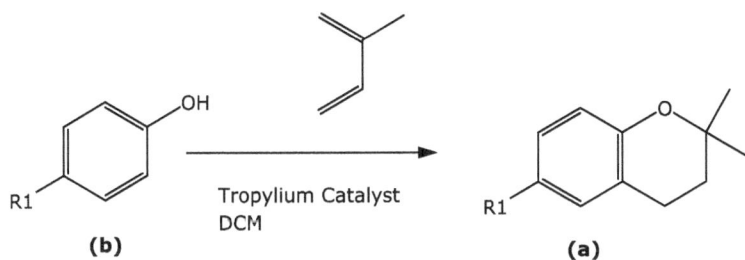

Fig. 6.1: Synthesis of 2,2-dimethylchromans (**a**) from phenols (**b**) using Vapourtec's R-series system.

Syrris's Asia flow system is a useful laboratory modular system, offering a wide variety of possible configurations, including photochemical and electrochemical units. Asia Manager software is used to program the equipment, allowing either single or multiple reactions to be automated. An example of its use is the fluorination of cyclic β-ketoesters (Fig. 6.2, **c**, for example) to give monofluorinated compounds (**d**) using N-fluorobenzenesulphonimide (NFSI) and a chiral copper catalyst [242]. A flow reaction was carried out using a microreactor module at 50 °C, the conditions being optimised to give high yields and enantioselectivities.

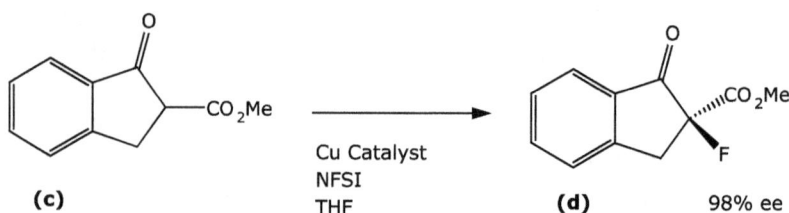

Fig. 6.2: Enantioselective fluorination of β-ketoesters (**c**), optimised using Syrris's Asia system.

The Asia system can be adapted for electrochemistry by incorporating the appropriate module. An example [243] is the continuous α-methoxylation of N-formylpyrrolidine (Fig. 6.3, **e**) to give 2-methoxy-N-formylpyrrolidine (**f**), The reaction takes place via anodic oxidation and would be impossible to carry out directly by non-electrochemical means.

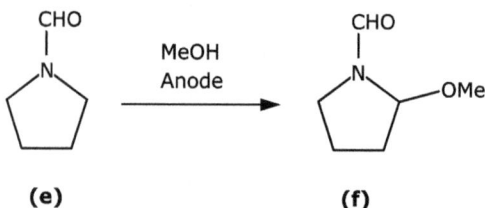

Fig. 6.3: Methoxylation of N-formylpyrrolidine (**e**) by automated flow electrochemistry.

ThalesNano produces the Phoenix Flow Reactor for laboratory-scale reactions, allowing high temperatures (up to 450 °C) and pressures (up to 200 bar) to be utilised. The reactor incorporates HPLC pumps and a control module. The company is also known for producing the H-Cube hydrogenation units, which allow for flow hydrogenations to be carried out without the need for a hydrogen cylinder, since the gas is produced in situ by the electrolysis of water. Sealed catalyst cartridges eliminate the need for hazardous catalyst filtration. An interesting example of its use is in studies on hydrogenation using scrapped catalytic converters [244], a variety of organic compounds being reduced in good yields.

Creaflow specialises in flow reactors, such as the HANU 15, that have a transparent window suitable for photochemistry. These reactors are available either as sole items or as part of an integrated modular system (MPDS EVO), allowing automated photochemistry. The HANU 15 has been used in combination with a pulsator to create an oscillatory flow reactor, which has been used in photochemical reactions with insoluble catalysts. For example, a Buchwald–Hartwig reaction, with Ni(II)Br$_2$·3H$_2$O and a graphitic carbon nitride (CN-OA-m) as catalysts rather than the usual palladium complex catalyst, was used to convert aryl bromides (such as that shown in Fig. 6.4, **g**) to amines (**h**) using a 460 nm LED light [245]. In general, flow reactions with insoluble solids have to be carefully optimised to avoid clogging.

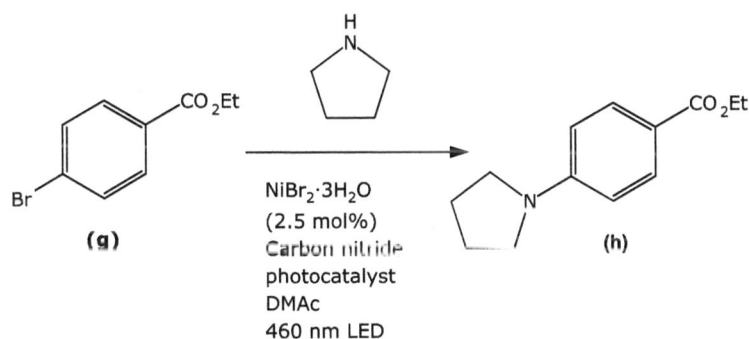

Fig. 6.4: Nickel-catalysed photochemical conversion of aryl bromide (**g**) to amine (**h**).

H.E.L's FlowCAT is a high-pressure flow system for catalytic reactions, allowing input flows of liquids and gases to be automatically controlled. The reactors can take pressures up to 100 bar, but options exist for a 200 bar rating if required. The system is now offered with H.E.L's new labCONSOL software, which builds on its WinISO software. Several reactions, such as hydrogenations, oxidations and carbonylations, can be carried out. For example, Steven Ley's group at Cambridge University used the FlowCAT system for the partial reduction of ethyl nicotinate (Fig. 6.5, **i**) to the olefinic compound (**j**) [246].

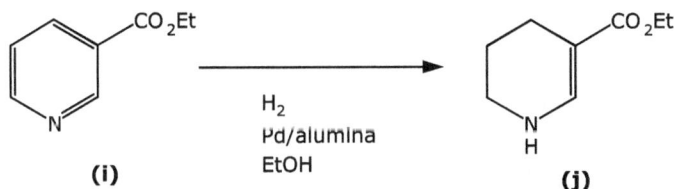

Fig. 6.5: Partial reduction of ethyl nicotinate (**i**) to (**j**) using the FlowCAT system.

Of course, it is quite possible to make your own automation systems, using a suitable computational device, usually a PC or laptop, and programming language. Another paper by the Ley group described the control of continuous reactions using the humble Raspberry Pi microcomputer programmed using Python [247], while a third described automated control using the internet, allowing remote alarms and control [248]. The control of a collection of sensors and devices via the internet is an example of the industrial internet of things (IIoT), which can be used to control not only production, but also the use of resources such as energy and water. LabVIEW graphical engineering software, from the US company National Instruments, has been used for reactor control [249, 250]; it is compatible with National Instruments' hardware and many third-party devices. Visual Basic has also been used for reactor control, despite this language being considered outdated by professional programmers.

Traditionally, the automated control of chemical reactions has involved so-called PID control [251]. Here the response, such as heating or cooling, depends on three factors: P (proportional term – speed of response, depending on how far you are from the set point), I (integral term – integration of the error over time, stopping the system from approaching the set point too slowly) and D (derivative term, an estimate of future error, stopping the value 'yo-yoing' up and down too much). In cases where such yo-yoing is not a problem in practice, only the P and I terms may be needed. PID controllers were used long before the modern computer existed and have successfully controlled vast numbers of processes. However, various forms of so-called fuzzy logic are now increasingly used for control purposes [252], offering advantages in some cases. As opposed to traditional (Boolean) logic that relies on 'true' and 'false' statements, 'fuzzy logic' involves concepts such as 'degrees of truth'. Learning strategies can also be used for control. For example, reinforcement learning directs a control system to continually improve based on the experience of previous runs [253]. Chemical engineers should have a thorough grasp of the various control algorithms, so it is good to consult them, particularly when changing from one control method to another.

6.2 High-throughput experimentation (HTE)

In the early stage of developing a process, you may want to screen a large number of catalysts or other reagents. Multiple solvent mixtures may also need to be screened for recrystallisations, particularly where polymorph screening is required. In order to save time, high-throughput experimentation (HTE) methods can be used [234, 254]. Many larger companies have a dedicated HTE group, with specialist equipment to carry out multiple small-scale reactions or recrystallisations. Systems often use 96-well plates or 96-vial set-ups, similar to those common in biological manipulations, with stirring typically carried out using vortex methods. Large multistirrers, such as

Thermo Scientific's Variomag Telesystem stirrer, which can give up to 60 stirring points, can also be used. Autosamplers allow HPLC to be carried out, so that the progress of the reactions can be followed, although in some cases other methods, such as SFC, GC, MS or changes in UV absorbance can be used.

The Junior system from Unchained Labs has a moving four-tip liquid dispenser above multiple vials. The latter can be vortexed, heated or cooled as required. In addition to vials, the Junior system can be fitted with up to eight small-scale reactors (referred to as 'Optimisation Sampling Reactors', OSRs), which allow reactions to be carried out under pressure- and temperature-controlled conditions, with regular sampling, allowing kinetic studies to be carried out. The system uses Unchained Labs' LEA software, which can both control the device and permit the importation of experimental designs.

The CM3 platform from Freeslate (now owned by Unchained Labs) allows automation using a variety of 'arm' (i.e. a robotic 'arm') and 'deck' configurations. Features include an integrated balance, a solid-dispensing system, a carousel for vials and eight high-temperature and high-pressure reactors. In addition, the arm can reach outside of the CM3 'deck' to access third-party equipment. Three independently controlled temperature zones are included, each of which can be used for multiple vial-scale reactions, allowing hundreds of reactions to be carried out per day. Again, Unchained Labs' LEA software is used. An example of the use of the CM3 is in the production of a range of transition metal catalysts for the dehydrative decarbonylation of hydrocinnamic acid (**k**) to styrene (**l**) using a variety of ligands and sources of metal [255] (Fig. 6.6). The most promising systems, which had nickel catalysts, were then applied to reactions with nonanoic acid to give 1-octene. The weak UV absorbance of the latter compounds made them less suitable for initial HTE compared to the aromatic species.

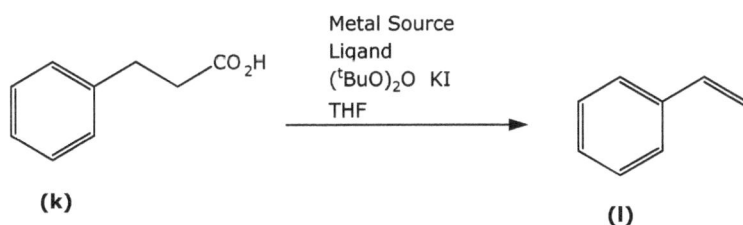

Fig. 6.6: Dehydrative decarbonylation using metal catalysts, optimised using the CM3 system.

Overall, the ubiquitous Suzuki–Miyaura coupling is the type of reaction most often optimised using HTE methods by pharmaceutical companies [234]. Here, a variety of palladium complexes (in some cases also nickel complexes), bases and solvents can be optimised to give suitable coupling systems (Fig. 6.7). A Suzuki–Miyaura HTE screening might involve a 96-vial system, covering 16 metal complexes, three bases and two solvents.

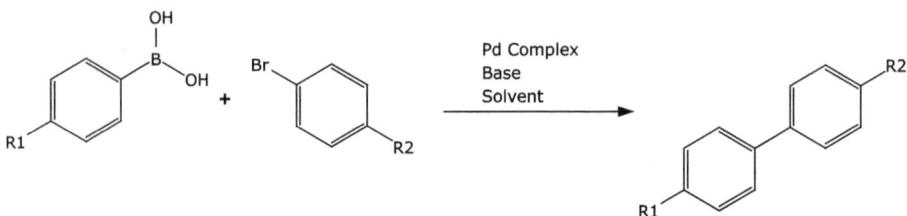

Fig. 6.7: Typical Suzuki–Miyaura coupling, with variations possible in Pd complex, base and solvent.

HTE enables a large number of reactions to be carried out simultaneously, but there is a danger that the analytical methods will become a 'bottleneck'. For HPLC, specialist parallel instruments or multiple injections on a single column can be used [256]. Modern UHPLC techniques or SFC can offer much shorter run times than traditional HPLC. In some cases, MS methods without chromatography, such as 'direct analysis in real time' can be used [257].

Although HTE is a useful method for finding promising catalysts or reagents, it should be emphasised that such small-scale reactions do not provide an accurate model for plant processes. The best catalyst or reagent may be missed because of poor stirring, surface effects, etc. It is strongly recommended to take at least two candidates forward for further examination on a larger scale, thus increasing your chances of finding a good system. Of course, HPLC %areas derived from HTE studies can be misleading, particularly if impurities are not sampled to the same extent as the product. In some companies, where one report leads on to another, it is not uncommon for '%area of product by HPLC' to end up as '%yield of product', which may prove very embarrassing if subsequent large-scale reactions show that the actual isolated yield is far lower.

6.3 Column automation

Preparative chromatography is a useful purification technique, although usually restricted, in practice, to relatively high-value products. Column optimisation is typically carried out on a laboratory scale, often after initial TLC investigations. However, the manual changing of fractions takes up much time, and staff can be freed for other tasks if this can be avoided by the use of automation. Column automation involves pumping in solvent from suitable reservoirs, detecting product (normally by UV) and collecting fractions. Many commercial automated fraction collectors tend to be suitable for small-scale preparative HPLC on a few grams. Cytiva's F9-R can be used in combination with Äkta preparative chromatography systems, being able to collect fractions up to 50 mL. Automatic peak recognition means that unwanted eluent can be diverted to waste. Buchi's C-660 system allows 250 mL tubes to be used and also features automatic peak recognition.

For larger-scale runs, Bio-Rad's NGC Fraction Collector can deal with bottles up to 250 mL, but the useful 'Prep Rack Adaptor' allows up to 80 collecting bottles of any size to be used with suitable connecting tubing. Bio-Rad's ChromLab software is used for control. Thermo Scientific's AFC-3000 automated fraction collector can be set up with various sizes of vials or tubes. A funnel rack with 21 positions can be used to direct chromatographic flow to containers of any size. The system uses Thermo Scientific's Chromeleon chromatographic software. Biotage's Isolera LS system is fully automated and can be used with collecting vessels up to 10 L in volume if a funnel rack is employed. It is recommended for up to 150 g input.

Automated fraction collectors tend to be somewhat expensive, so smaller companies may wish to construct their own alternatives. Even 'toy' systems can be used: the 'Lego Mindstorms' robotic system has been employed to produce an inexpensive 'home-made' fraction collector [258]. Three-dimensional printing has also been used to produce an in-house fraction collector for around €100 [259].

6.4 Peripheral equipment

Automated systems require suitable peripheral equipment, such as thermocouples, pumps and level gauges. Commercial automated systems typically contain their own peripheral equipment, while some also allow the software to control other manufacturers' systems. However, such third-party peripherals have to be compatible; for instance, they may have to be 'RS-232-enabled' (RS-232 is a data transmission standard). If you are making your own automated system, care must be taken to ensure that the peripheral equipment is compatible. Many modern thermocouple data loggers come with USB connections and their own software, usually allowing a straightforward interface with the controlling computer.

Peripheral equipment must be able to withstand chemical degradation. Metal temperature probes are obviously vulnerable to corrosion, but give a quicker response than PTFE-coated sensors, which can be important in highly exothermic reactions. Redox electrodes are often used to control continuous oxidations, halogenations or other reactions where the amount of reagent needs to be limited. Care must be taken to find compatible electrode materials: graphite, platinum, silver, gold, tantalum and tungsten are among those that have been used. For example, various azo-coupling reactions were carried out under redox control using a tungsten electrode (Fig. 6.8) [260]. Both redox and pH measurements are affected by temperature, so if this changes significantly during the reaction, corrections may be needed.

Peristaltic pumps are commonly employed for automated reactions, but can't be used with ordinary PVC or PTFE tubing, so silicone tubing is often used instead. This is produced using either platinum curing or peroxide curing. The former is more

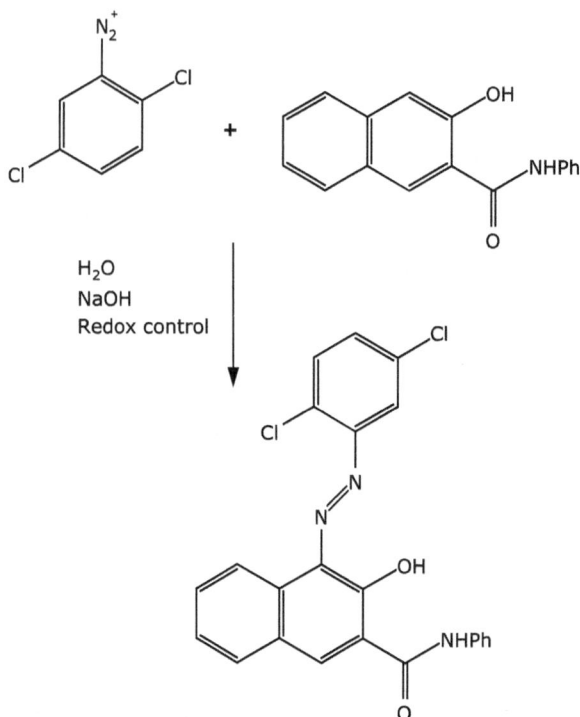

Fig. 6.8: Azo-coupling under redox control using a tungsten electrode.

expensive, but the latter, although suitable for most purposes, can sometimes leach by-products of the curing process (organic acids). All silicone tubing can be damaged by a number of liquids, particularly strong acids, hydrocarbons and ketones (e.g. acetone). Tygon tubing is typically more resistant to hydrocarbons than silicone tubing: it is a polymer mixture, available in various grades with differing compositions. Viton rubber is a fluoroelastomer-based material that gives tubing with good chemical resistance, including greater acid resistance than silicone. For example, it has been used for the output from bromination reactions [261]. A number of manufacturers offer tubing compatibility guides on their websites, but it is worth carrying out stability tests with various types of tubing if problems are at all likely.

Since silicone tubing is incompatible with many substances and PTFE tubing gives far wider chemical compatibility, it is often necessary to join the two together for continuous reactions. Joining different types of tubing can lead to leaks and is best done using suitable adaptors and clips. Simply squeezing one tube into another is not advised, even if no leaks are found on initial testing. Heat shrink tubing can be useful in some cases for joining tubing to adaptors, glass tubing, etc. Figure 6.9 is a schematic diagram showing the use of a peristaltic pump in a semi-batch laboratory chlorination reaction, with chlorine being formed in situ in controlled quantities. The pump adds 35% hydrogen peroxide to a mixture of hydrochloric acid and substrate, first going through silicone tubing and then through PTFE tubing. Adding the acid to the peroxide would not be possible using silicone tubing, since strong acids tend to cause degradation. The addition rate can be controlled using redox electrodes or a UV/visible device. The pump can also be programmed to automatically switch off should the cooling fail to control the temperature.

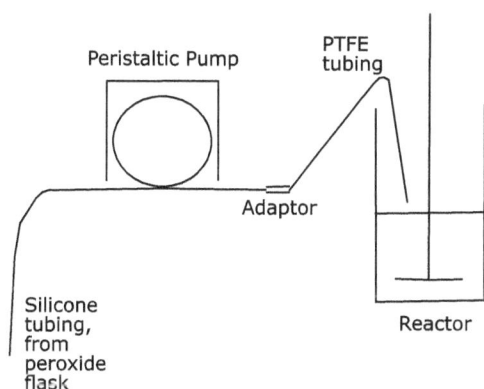

Fig. 6.9: Use of peristaltic pump for controlled addition of hydrogen peroxide.

Equipment in automated reactors may be affected by fouling, with a build-up of solids or tars causing inaccurate temperature, pH or redox readings. Online analysis, such as UV or NIR, may also be seriously affected. Blockages can affect pumps, tubing and vents (in the latter case, from sublimation and subsequent solid deposition, which can occur with substances such as ammonium chloride). Such occurrences may pass unnoticed, particularly if they take place in the middle of the night, which no doubt they will. It is important that automatic systems are programmed to safely switch off if unusual conditions are detected. An extra 'cut-out' thermostat may stop overheating in the event of the control thermostat failing. A pressure relief valve is required if over-pressurisation is a possibility. In addition, remote alarms and web-cam surveillance can be useful methods for checking on reactions when nobody is present.

6.5 Choosing automated systems – present and future

There are many automated systems on the market, and constraints of space mean that not all can be mentioned in this book. Careful consideration is needed before any are purchased. The wrong decision may end up as an expensive white elephant, which only makes an appearance on open days or when managers from other companies pay a visit. It is important to objectively evaluate what you really need. If the only real requirement is to start a pump and stirrer at 3 am, a simple mains timer plug may be all that's called for.

If a proper automated system is required, it is possible to produce one 'in-house', provided there's somebody with the necessary skill to write a control program. However, if finances permit, it normally saves a lot of time if a commercial system is chosen, where all the various hardware and software bugs have (in most cases) been sorted out previously. Make sure that any such system has the particular features you require, including the ability to deal with different sizes of reaction. For example, with a continuous automated reactor, you will probably want to replace syringe pumps with peristaltic pumps as you scale up, while also possibly needing to increase the reactor dimensions and cooling efficiency.

Modular systems tend to work out cheaper in the long term, since different modules can be purchased as new projects come up, rather than a new system having to be acquired. For example, some flow reactors offer both photochemical and electrochemical modules. Although you may not need these at present, you never know what the future may bring. Photochemistry is increasingly being used, particularly for photoredox catalysis, so expect to have to scale up such reactions at some point. In the past, electrochemistry tended to be the preserve of a few academic and industrial groups, but its usefulness is now beginning to be more widely appreciated. It may well be that before the end of the century, many reactions will be carried out automatically in continuous mode, by adding or subtracting electrons, while the steam-heated batch reactor will seem as quaint as the steam locomotive does today.

Beyond process development, much effort has been expended on producing fully automated robotic systems for synthesis [262]. For example, the 'Chemputer' system devised by Lee Cronin's group at the University of Glasgow can carry out numerous different reactions in succession, without manual intervention [263]. Initially, such systems will probably be used mainly by research departments but will have applications in process development for producing impurity standards, etc. Ultimately, we may see robotic systems that can not only carry out a range of chemical reactions, but also optimise them, without the need for human input. However, it will be many years before process development chemists are no longer required, since the intricacies of scale-up are much harder to program compared to simply producing an uncomplicated molecule on a small scale.

Chapter 7
Environmental and toxicology issues

7.1 Green chemistry, waste and solvents

Green chemistry involves devising processes with the minimum adverse environmental effects, although there are, in practice, a number of different ways this can be assessed. One obvious aim is to minimise waste. However, not all waste is equal: a tonne of sodium chloride is obviously less of an environmental hazard than a tonne of mercury chloride. Modern industrial practice in organic chemistry is to avoid the use of heavy metals except in catalytic quantities, in order to eliminate such waste disposal challenges. One of the main principles of green chemistry is to avoid the generation of environmentally damaging waste rather than try to deal with it after it has been formed.

Waste streams from processes include aqueous waste, typically containing inorganic ionic species, such as acids, alkalis or salts. Acidic waste streams are often neutralised 'in-house' with sodium hydroxide before being sent for disposal. In addition, there is aqueous waste from gas scrubbers: a common mixture would be sodium hydroxide and sodium chloride (from a process that gives off hydrogen chloride). In some cases, aqueous waste may have a possible secondary use. In theory, aqueous nitrate or phosphate solutions could be used as fertiliser, although in practice regulatory restraints may make such usage unfeasible. Even if it presents no environmental danger, proving the point may be too costly to be worthwhile, although regulatory progress on such issues may occur in the future. In most companies, aqueous waste is either sent off site for disposal or treated at an in-house wastewater treatment plant to make it suitable for discharge to waterways. However, large quantities of aqueous waste containing significant amounts of toxic water-soluble polar aprotic solvents, such as DMF, can be difficult to dispose of. Such solvents are best avoided, if possible, particularly in larger-scale production.

Many processes give rise to organic solvent waste, which typically contains organic by-products from the process (if it contains a significant amount of product, your process needs a rethink). Solvent recycling reduces the amount of waste that needs to be sent for disposal, as well as saving money. If not recycled, organic solvent waste is typically sent for incineration. Chlorinated solvents can be particularly harmful to the environment, contaminating groundwater [264] and adversely affecting a range of species, including humans. Incineration can give rise to environmentally damaging and toxic chlorinated dibenzodioxins and benzofurans, unless the conditions are carefully controlled so that a consistently high temperature is maintained. Some chlorinated solvents, such as carbon tetrachloride, are banned or restricted due to their capacity to destroy the ozone layer. Chlorinated solvents are thus avoided in modern process development, unless there are no viable alternatives.

https://doi.org/10.1515/9783110717877-008

As mentioned in Chapter 1, ionic liquids are non-volatile, so are considered environmentally friendly since losses to the atmosphere are negligible. However, this property is a two-edged sword, since it means that recycling via distillation is impossible. In addition, ionic liquids usually require many more steps in their synthesis, compared to simple, common solvents, and thus require the use of more energy in their production [265].

Solid waste, often inorganic salts or filter aids, may also be removed by filtration during a process, typically going to a suitable landfill site. Spent heavy metal catalysts on filter aids should be recycled, and some suppliers offer a recycling service. Solid organic waste from stoichiometric reactions, such as triphenylphosphine oxide from Wittig reactions, can be costly to dispose of. Various processes have been proposed for the reduction of this oxide back to triphenylphosphine [266, 267], although currently these are rarely implemented.

While liquid and solid waste can be dealt with in an appropriate manner, typically by recycling, incineration or landfill, gaseous emissions present more of a challenge. As detailed in Chapter 2, scrubbers can be used to remove many gaseous by-products from plant processes – sodium hydroxide scrubbers removing acidic gases, acidic scrubbers removing basic gases such as ammonia, and organic solvent scrubbers removing neutral organic molecules. Gas monitoring may be needed to ensure that scrubbers are working effectively in practice.

The literature is full of 'green methods' for synthesis. Some are excellent, while many are either wholly impractical or have no advantage over existing methods. There is little point eliminating all organic solvent from a reaction if large quantities are then needed for the work-up. Again, running reactions without solvent, say by grinding a solid, or using microwaves on a starting material absorbed into a solid support, may only be possible on a laboratory scale for exothermic reactions, since the solvent and vessel cooling in the equivalent conventional reaction play an essential role in temperature control, reaction safety and reproducibility.

A bewildering array of metrics is used to describe the environmental effects of a process and its sustainability, some being of more utility than others. It is useful to consider the so-called 'atom economy' of a process, a concept first introduced by Barry Trost [268]. This is a measure of how many of the atoms going into the process remain in the final product. Figure 7.1 shows the reaction of cyclohexene (**a**) with 3-chloroperoxybenzoic acid (MCPBA, **b**) to give cyclohexene oxide (**c**) and 3-chlorobenzoic acid (**d**). The atom economy is given by the molar mass of the product (**c**) divided by the sum of the molar masses for the input molecules (**a** and **b**) expressed as a percentage. As mentioned in Chapter 1, only one out of 16 atoms in MCPBA is incorporated into the product. However, the reaction can also be carried out with hydrogen peroxide and a peroxotungstate catalyst [269], giving a higher atom economy.

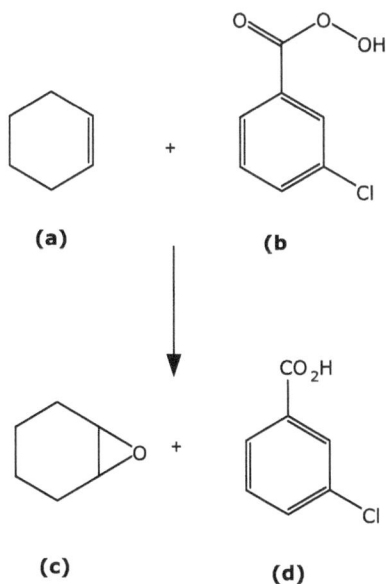

Fig. 7.1: Oxidation of cyclohexene (**a**) with MCPBA (**b**) to give cyclohexene oxide (**c**) and 3-chlorobenzoic acid (**d**).

The atom economies of the two oxidation methods are calculated below:

Atom economy (MCPBA): molar mass (**c**) x 100/(molar mass (**a**) + molar mass (**b**)) = (98.1 x 100)/(82.1 + 172.6) = 38.5%

Atom economy (hydrogen peroxide): molar mass (**c**) x 100/(molar mass (**a**) + molar mass H_2O_2) = (98.1 x 100)/(82.1 + 34.0) = 84.5%

The above calculation is an approximation, since a small adjustment should be put in place for the catalyst, since it is not used for ever, while in practice an excess of peroxide would be used. However, the overall message is clear: bulky stoichiometric reagents such as MCPBA waste many atoms. It is possible to recover 3-chloroperbenzoic acid and oxidise it back to MCPBA, but losses would still make the atom economy of the overall process worse than one using hydrogen peroxide.

Another measure of the 'greenness' is the E factor [270], initially introduced by Roger Sheldon, which is the mass of waste divided by the mass of product. 'Waste' in this context includes solvents, salts in aqueous streams (but not water itself) and gaseous emissions. A high E factor is a clear warning sign that a process needs to be improved, so it is worth calculating when comparing different processes and routes. However, the E factor does not differentiate between the different types of waste: a kilo of sodium chloride counts the same as a kilo of mercury chloride, so it is not useful for comparisons between different processes when the environmental impact of the waste differs greatly. This problem can be solved by using the so-called environmental quotient, EQ, which is the product of E and an 'unfriendliness' quotient, Q, which is assigned based on quantities such as toxicity or whether the waste can be readily recycled. The values of Q are small for relatively benign waste, such as sodium chloride, but much higher for heavy metal salts, but will differ somewhat, depending on the assumptions of the person carrying out the assessment.

In theory, a full life cycle assessment (LCA) can be carried out for any particular product, but the lack of readily available data usually makes this difficult during process development [271]. LCA requires a full assessment of the impacts of a product's formation, use and eventual disposal. A useful tutorial review covering LCA in the chemical industry has been recently published [272]. In practice, LCAs tend to be carried out on high-volume existing products, where supply chains and environmental fate are well established.

A more easily applied metric is process mass intensity (PMI), which is the total mass of raw material input divided by the mass of product produced. The inputs include solvents and water. PMI is typically applied to an overall process, rather than a single step, and is often used in the pharmaceutical industry [273]. As an example of PMI, we can consider a simple acid-catalysed reaction. An initial process (A in Tab. 7.1) used 10 kg of solvent and 0.5 kg of acid catalyst per kg of starting material and gave 0.9 kg of product. Three 3 kg water washes were used for work-up. A second process (B in Tab. 7.1) gave the same yield, used the same amount of solvent, but 0.5 kg of a different acid catalyst, which only required a single 3 kg water wash.

Tab. 7.1: PMI values for two processes, differing in the amount of water used.

	Process A	Process B
Starting material per kg product	1/0.9 = 1.1111	1/0.9 = 1.1111
Solvent per kg product	10/0.9 = 11.1111	10/0.9 = 11.1111
Acid catalyst per kg product	0.5/0.9 = 0.5555	0.5/0.9 = 0.5555
Water per kg product	9/0.9 = 10	3/0.9 = 3.3333
PMI	22.8	16.1

Thus, switching to process B has decreased the PMI value by 29%. Clearly, assuming there are no significant price, safety, or environmental issues in switching to process B, it is to be preferred. However, the advantages of lowering the amount of water used will vary greatly, depending on the location of the plant. In an arid region, great care needs to be taken to use the minimum of water, while in a part of the world where it is nearly always raining, water usage may be much less of an issue.

PMI focuses on overall sustainability rather than simply on waste and is useful for comparing different routes to the same product: a route with a particularly high PMI is unlikely to be sustainable. However, PMI can be misleading [274] when comparing an optimised reaction, where concentrations are likely to be high and washes minimised, with a 'research method', where the reaction may be run at a much greater dilution than necessary and the work-up may contain several unnecessary washes, etc. Even when comparing 'like with like', PMI should not be used as a sole metric, since other considerations, such as the environmental effects of waste and the sustainability of inputs, need to be taken into account.

7.2 Carbon footprint and long-term sustainability

The world is facing a climate crisis due to increasing amount of carbon dioxide and other greenhouse gases in the atmosphere [275]. Avoiding the worse effects of climate change will involve huge cuts to the amounts of fossil fuels used, eventually leading to net zero greenhouse gas emissions. Like all other industries, the chemical industry will have to greatly reduce its carbon footprint. Although the bulk of greenhouse gas emissions come from the petrochemical industry, rather than fine chemicals and pharmaceuticals, all parts of the industry will have to consider how to minimise their carbon footprint. The carbon footprint of the pharmaceutical industry is not negligible, but does tend to vary significantly between companies, suggesting room for improvement in some cases using existing technologies [276]. Although there does not seem to have been a comparable exercise to date for other sectors, such as agrochemicals or perfumery compounds, it seems likely that similar results would be obtained.

Even if electricity generated via renewable resources is used in the future, it is likely to be relatively expensive, so progress is needed in minimising its use. Although much avoidable energy waste is beyond the scope of process development, such as leaky and ineffective steam and compressed air systems on plants, it is still necessary to develop chemical processes that minimise energy use, while still maintaining yields and minimising waste. A significant amount of energy is expended at present in heating up or cooling down reaction mixtures and distilling solvents. Some distillations could in theory be replaced by membrane separations, but progress has been slow since most membranes are corroded by organic solvents. Recent developments of more resistant membranes for the separation of organic compounds offer promise that this technology could be widely used in place of distillations and extractions [277]. For example, a polymer/metal oxide membrane gave 53% rejection for diisopropylbenzene and 92% rejection for triisopropylbenzene, showing useful size discrimination [278].

Both photochemistry and electrochemistry are promising technologies that can be used in certain cases to transform molecules without heating them. Photoredox technologies allow a variety of reactions to be carried out at ambient temperatures, avoiding stoichiometric reagents and generating little waste [279]. In some cases, reactions can be carried out that would be impossible thermally, cutting down on the number of synthetic steps and the associated energy use and waste. An example is the use of methane, ethane, propane and isobutane as alkylation agents under photochemical flow conditions with a tungsten photocatalyst [280] described by a multi-centre team led by Timothy Noël from Einhoven University of Technology, Netherlands. Alkylation was carried out on electron-deficient alkene substrates, such as that shown in Fig. 7.2 (**e**), which was converted to compounds such as the ethylation product (**f**) in the presence of a TBADT (tetrabutylammoniumdecatungstate) catalyst. A conventional route involving conversion of the alkane to an alkyl halide, metalation and reaction with the alkene would be much less green, giving a higher volume of waste and involving at least two separate reactors and difficult separations (for instance, the halogenation is unlikely to be totally selective).

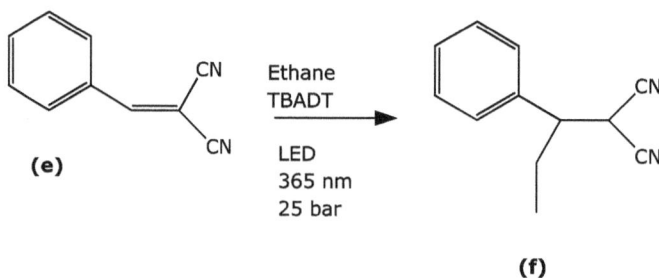

Fig. 7.2: Ethylation of alkene (e) to give (f).

Thus, while the use of innovative technologies, the optimisation of plant energy use and the employment of electricity generated from renewal resources can all lower the carbon footprint of chemical production, a more fundamental problem remains. The bulk of the carbon atoms used in pharmaceutical and agrochemical products are derived from the petrochemical industry, ultimately from fossil fuels such as crude oil (the main source of carbon), natural gas or coal. However, as climate change becomes more severe, it is likely that governments will, sooner or later, adopt policies of leaving fossil fuels in the ground. As the world pivots away from fossil fuels, it follows that the feedstock availability for the chemical industry will become more uncertain. Even if fossil fuel production continues at current levels (which would no doubt lead to catastrophic climate change), eventually easily recoverable fossil fuels will become scarcer as they are used up, so a transition to renewable feedstocks will be required at some point.

In the future, apart from legacy production from fossil fuels, there are likely to be three main sources of carbon: biomass (such as ethanol from crops), carbon dioxide from gas emissions (carbon capture and utilisation) and waste materials (such as methane generated from waste). Radical changes will be needed throughout the chemical industry as fossil fuels are phased out.

Biomass is used on a large scale to produce ethanol, which is mainly employed in petrol. The EU is currently moving from petrol containing 5% ethanol to a grade containing 10%, while elsewhere similar uses of ethanol are common. Apart from ethanol, a relatively small number of other chemicals, such as the furfural mentioned in Chapter 1, are currently produced from biomass. It is considered that integrated biorefineries are the most effective method of producing useful chemicals from biomass [281], although further work is needed to improve conversions in order to make biorefineries competitive. It has been suggested that the key to increasing the proportion of chemicals produced from biomass is to focus to a greater extent on farmers, making sure they are properly rewarded for producing the necessary inputs [282].

It is important to bear in mind that just because a particular chemical occurs in a plant or microorganism, it does not mean that it is a viable source. In particular, if the concentration is low and vast amounts of solvent would be needed for extraction and preparative chromatography, the isolation process can become both costly and environmentally damaging. Genetic modification is a possible way of increasing yields. GM microorganisms are widely used to make therapeutic proteins, and such techniques have now been extended to smaller molecules. For example, GM yeast has been used on a laboratory scale to produce opioids [283].

The production of chemicals by the reduction of carbon dioxide is currently a topic of great interest for academic researchers [284], although it is not usually run on an industrial scale. Carbon dioxide is a promising feedstock, which is potentially available in very large quantities from carbon capture in fossil fuel power plants, and its use can avoid its adding to climate change by being discharged to the atmosphere. Even if the burning of fossil fuels eventually ceases, carbon dioxide should still be available from the burning of biomass and also from fermentation processes.

Electrocatalytic reduction of carbon dioxide using electricity from renewable energy is a promising method, which can produce both C1 (carbon monoxide, formic acid, methanol and methane) and C2 (ethylene and ethanol) products [285, 286]. Photochemical reduction using catalysts, either transition metal-based or organocatalysts, can produce formic acid [287] and other C1 species. Carbon dioxide can also be incorporated into molecules by reactions with diols or olefins to give cyclic carbonates [288]. Figure 7.3 shows some of the main reactions of carbon dioxide that give non-polymeric molecules.

The world is awash with plastic waste, making it is a promising source of feedstock materials. Polyesters can be hydrolysed either chemically or enzymatically [289], and the resulting monomers reused to make fresh polymer or transformed to other molecules. Obtaining useful chemicals from waste polythene is more difficult, since pyrolysis gives a complex mixture of products. It has recently been shown by Susannah Scott's group at the University of California that polythene can be catalytically converted to long-chain alkyl aromatics, which are possible detergent precursors [290]. Waste polystyrene can now be converted back to styrene monomer on an industrial scale by pyrolysis and has also been broken down at near-ambient temperatures on a laboratory scale [291]. It is likely that increasingly better methods will be found to produce useful chemicals from polymer waste, and the main challenge in

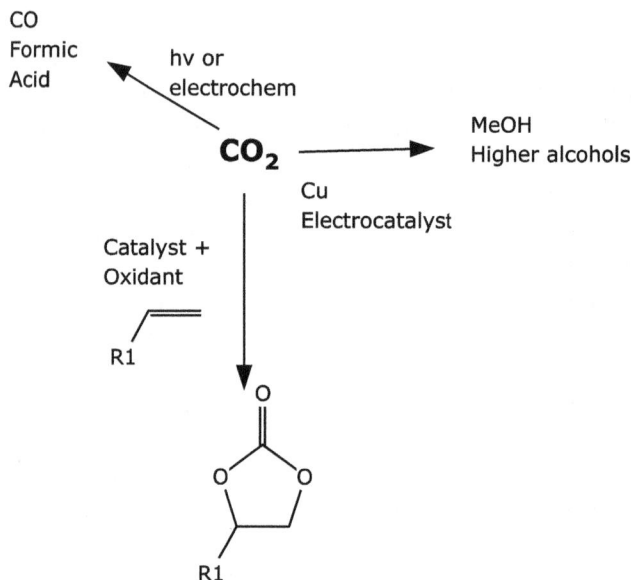

Fig. 7.3: Some reactions of carbon dioxide to produce feedstock chemicals.

the future may be the effective collection and separation of the plastic streams. Waste plastic needs to be utilised by the chemical industry rather than being stuck in landfill or floating around the oceans.

The sustainable industrial chemistry of the future will look different to the petrochemical-based chemistry we are used to. Changes to feedstocks will result in changes to prices, benefitting processes based on substances that are accessible from biomass, carbon dioxide or waste materials. On the other hand, relying on compounds that are only readily available via petrochemicals may make a process become less competitive as the years go by.

7.3 Scavengers

Transition metals, many with a relatively high toxicity and environmental impact, are used as catalysts in a wide variety of chemical processes. Usually the bulk of the metal can be removed by filtration, often using a suitable filter aid. However, small amounts of metal, maybe complexed and kept in solution by organic ligands, can often remain in product streams. Since the limits on heavy metals are very low in most regulatory regimes, sometimes <1ppm in the final product (for pharmaceuticals, the exact limits depend on the dosage), scavengers, as previously mentioned in Chapter 3, may be needed to remove sufficient metal to meet the requirements in force.

Ethylenediamine derivatives and related compounds chelate heavy metals. Eisai chemists demonstrated the removal of palladium residues from the crude product from a Suzuki–Miyaura coupling using polymer-supported ethylenediamines [292]. The initial palladium levels were 2,000–3,000 ppm, being reduced to 100–300 ppm. Subsequent salt formation from the crude product further reduced the palladium levels to less than 10 ppm.

Sulphur-containing species can also bind strongly to metals. Silica scavengers modified with multidentate sulphur, as shown in Fig. 7.4, were used to remove palladium down to less than 5 ppm [293]. Experimental designs have been used to optimise the best conditions for palladium removal with one of these scavengers [294]. As mentioned in Chapter 3, potassium isopropyl xanthate is a particularly useful S-containing scavenger, able to reduce Pd levels down to 1 ppm from around 50 ppm. Even greater reductions can occur when it is used in combination with iodine, a synergistic effect existing between them [143]. Metals can also remove each other; for example, iron powder has been used as a copper scavenger in flow reactions [295].

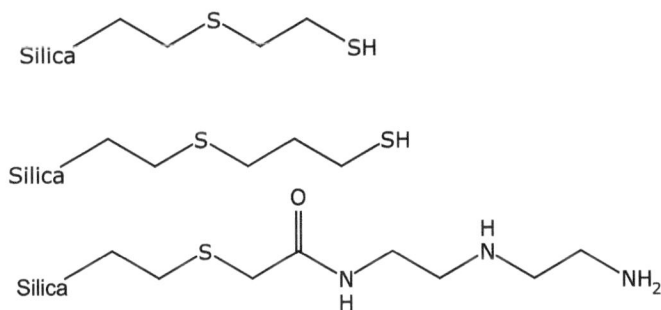

Fig. 7.4: Sulphur-containing palladium scavengers on silica supports.

A wide range of scavenger resins are commercially available. They are often used in combichem, but also have applications for large-scale synthesis. Scavenger resins can be used in packed columns in continuous flow reactions, removing excess reagents or by-products from reaction mixtures as an alternative to a conventional work-up. Various types of scavenger resins can be used: for example, acidic resins can remove bases, basic resins remove acids and benzylamine resins can remove unwanted aldeyhydes [142, 296, 297]. Apart from the safety and control advantages of flow reactions, such specific scavengers may be used to isolate and reuse valuable reagents and by-products that might be hard to recover from a normal batch work-up.

7.4 Mutagenic residues in products

The intrinsic toxicity of products to their users is outside the scope of process development, but we can at least ensure we do not add to it with unwanted toxic residues. A variety of mutagenic impurities can contaminate products, and strict limits are in force, the impurity levels usually being designated as critical quality attributes. Much effort has been expended in minimising the levels of mutagenic impurities in pharmaceuticals [298], but they can also be an issue in agrochemicals [299, 300]. A historic example is the contamination of herbicides with chlorinated dioxins, which were responsible for the appalling health effects of the 'Agent Orange' used in the Vietnam War. With regard to pharmaceuticals, until recently much of the literature used the term 'genotoxic', but since the implementation of the ICH M7 (R1) international guidelines [301], more precisely defined terms such as 'mutagen' have been preferred. The guidelines give details as to how to calculate acceptable daily intakes for mutagens. Acceptable levels depend on the dose and whether the pharmaceutical is taken over a long duration or not.

Traces of mutagens can be a serious problem for industry, resulting in products being recalled and production halted, while process development chemists frantically try to lower the levels of the offending impurities. Particular problems have occurred with compounds that do not readily show up on HPLC. Mutagenic N-nitroso compounds have led to severe difficulties in a number of cases. Ranitidine (Zantac, Tritec) is a commonly used drug that reduces excess stomach acid (H_2 receptor antagonist). However, ranitidine (Fig. 7.5, **g**) can be contaminated by the probable human carcinogen, N-nitroso-dimethylamine (NDMA, **h**).

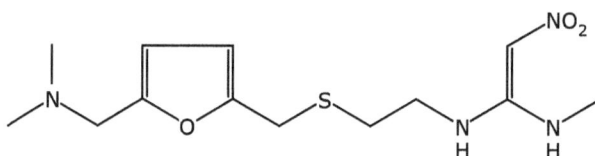

(g)

(h)

Fig. 7.5: Ranitidine (**g**) and NDMA impurity (**h**).

On the 1 April 2020, the FDA announced the withdrawal of all ranitidine, both prescription and over-the-counter (OTC), due to traces of NDMA increasing over time in some samples. An investigation by GSK scientists showed that the NDMA arose from the slow decomposition of the ranitidine molecule, rather than being formed from impurities [302]. The formation of NDMA was accelerated by high temperatures and humidity.

Another NDMA contamination problem arose with the hypertension drug valsartan (Fig. 7.6), when material from some manufacturers was found to be contaminated with low amounts of NDMA. Subsequently, some batches of valsartan were also found to be contaminated with the related impurity *N*-nitroso-diethylamine (NDEA). Recalls of valsartan led to shortages. *N*-nitroso impurities were later found in other 'sartan' products.

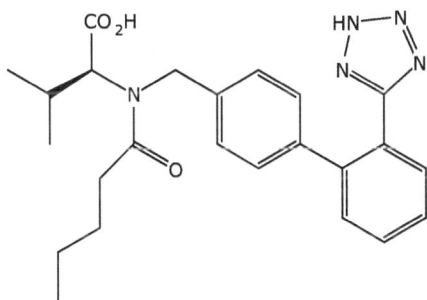

Fig. 7.6: Valsartan, a hypertension pharmaceutical contaminated with NDMA.

It has been suggested that the combination of DMF and sodium nitrite used in the step where the tetrazole ring is formed caused the formation of NDMA [303]. DMF can form traces of dimethylamine, which can then react with sodium nitrite to give NDMA. Figure 7.7 shows the formation of valsartan benzyl ester (**i**) from the nitrile intermediate (**j**). The sodium nitrite is used to destroy excess sodium azide, which needs to be removed, since it is explosive and can form toxic hydrazoic acid on acidification. The original valsartan process used tributyltin azide, but neither sodium nitrite nor DMF, so it was thought likely that the contamination occurred due to a process change. The whole episode shows the importance of thorough analytical testing whenever such changes are introduced.

The recalls associated with sartans and ranitidine have made industrial chemists much more aware of the possibilities for the formation of *N*-nitroso compounds and for the need for accurate analytical methods for these contaminants. Certainly, any process using sodium nitrite should be looked at carefully. Apart from product stability and reaction by-products, other steps in processes need to be considered. It is possible that levels of NDMA were increased, in some cases, due to solvent recycling [304].

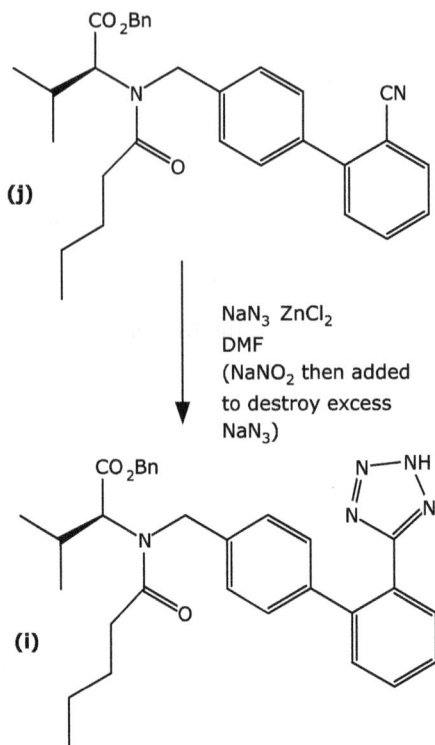

Fig. 7.7: Tetrazole ring formation: valsartan benzyl ester (i) is made from nitrile (j).

Apart from *N*-nitroso compounds, another area of concern is mesylate and tosylate esters, such as methyl mesylate and ethyl mesylate. These compounds are mutagenic alkylating agents that may arise during synthesis, since the mesyl group is commonly used as a leaving group, being displaced by nucleophiles. A number of drug substances are sold as the mesylate or tosylate salt, and the likelihood of alkyl esters being formed during the drug formation process has been a cause for concern. A recall was put in place in 2007 for the HIV drug nelfinavir (Viracept), which is a mesylate salt, because of contamination with ethyl methanesulphonate, which was shown to have formed from reactions between methanesulphonic acid and traces of ethanol (used for cleaning) in a holding tank [305]. It seems that simply drying the holding tank and lines after the ethanol clean would have prevented the issue arising. Since then, routine testing for alkyl sulphonates has generally become a regulatory requirement for products that are sulphonate salts, although it has been postulated the fears of alkyl sulphonates forming are unwarranted in most cases of salt formation [306].

Mutagenic epoxides, which act as alkylating agents, can contaminate products. A 1980 paper [307] describes the contamination of β-blockers, such as that shown in Fig. 7.8 (**k**), by the epoxide precursor (**l**). Levels of about 1% of the epoxides were detected by HPLC. The contamination of cardiovascular drugs, which may be taken for decades, is a particular danger.

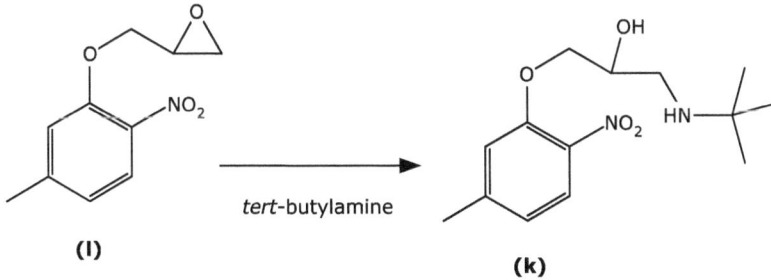

Fig. 7.8: Mutagenic epoxide precursor (**l**) for β-blocker (**k**).

Mutagenic epoxides can be formed by the nematicide 1,3-dichloropropene. The compound gave a positive reading in bacterial mutagenicity assays, but this was found to be due to a mixture of epoxide impurities [308]. Removal of the trace impurities using silica gel chromatography gave 1,3-dichloropropene that was inactive in the particular assay described. This agrochemical is no longer allowed to be used in the EU or the UK, but at the time of writing is still sold in many countries, including the USA.

Another potent alkylating agent and carcinogen is bis(chloromethyl)ether, $(ClCH_2)_2O$. This compound and related chloromethyl compounds can be formed inadvertently. For example, the reaction of formaldehyde and hydrogen chloride in moist air can form traces of this ether [309]. In general, alkylating agents are one of the most widespread classes of mutagens: they include alkyl sulphonates, dimethyl sulphate, alkyl phosphonates, epoxides and some alkyl halides.

Hydroxylamine and some of its derivatives are also mutagenic and may be formed by the breakdown of larger compounds. Hydroxylamine does not show up under typical HPLC with UV detection and GC conditions, although LC/MS can be used [310]. Hydrazine and its alkyl derivatives pose similar challenges, many being mutagenic and also hard to detect by conventional means [311].

Another class of carcinogenic and environmentally harmful chemicals are polychlorinated biphenyls (PCBs). These can be formed as an unwanted by-product of Suzuki–Miyaura reactions with substrates such bromodichlorobenzenes or dichlorophenylboronic acids, where, in addition to the desired coupling reaction, the molecule can undergo some coupling with itself to give traces of PCBs (Fig. 7.9). Careful optimisation of the reaction conditions is needed to minimise their formation [312].

Aromatic amines are a large class of potential contaminants, being intermediates for many pharmaceuticals and agrochemicals. Carcinogenic o-toluidine (2-methylaniline) is

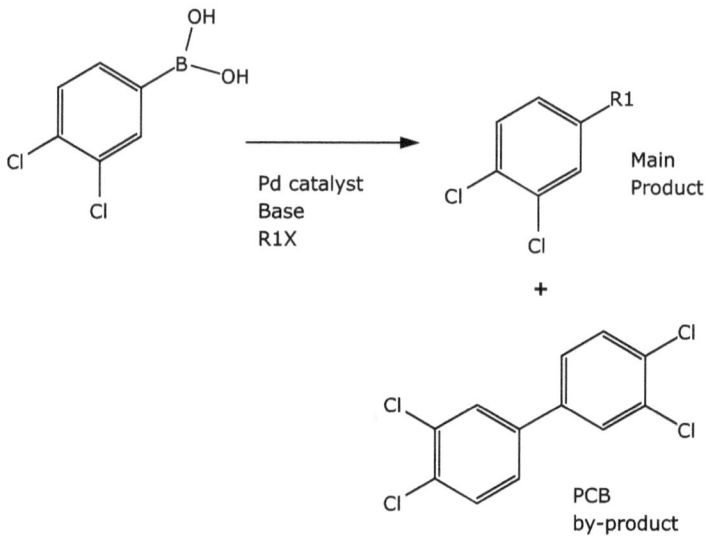

Fig. 7.9: Possible formation of PCBs as Suzuki–Miyaura reaction by-products.

a particularly dangerous potential contaminant [313]. Heavy metal impurities may also be a problem, and they may pass unnoticed if the metals are used to make starting materials [314]. Just because they aren't in 'your' process doesn't mean they aren't there.

Following early work by Ashby and Tennant [315] and subsequent refinements, several structural alert programs for potential mutagens are available. These can flag up potential mutagens by means of structure–activity relationships. Fragment-based methods are common, in which the bonds of the target molecule are broken and the resulting fragments evaluated. Various approaches and programs have been compared and evaluated [316, 317]. Such approaches save time and money, since potential impurities can be quickly checked, even if they are only hypothetical and no actual sample exists. The main classes of mutagens and carcinogens that may be found in pharmaceuticals and agrochemicals are listed in Tab. 7.2.

Tab. 7.2: Main classes of contaminating mutagens and carcinogens.

Class	Examples	Refs
Alkylating agents	Alkyl sulphonates, dimethyl sulphate, alkyl phosphonates, epoxides, lactones (some), alkyl halides (some)	305, 306, 307, 308, 309
N-Nitroso compounds	NDMA, NDEA	302, 303, 304

Tab. 7.2 (continued)

Class	Examples	Refs
Hydroxylamine and derivatives	Usually unsubstituted hydroxylamine	310
Hydrazine and derivatives	Free hydrazine, alkyl hydrazines, hydrazones	311
Polychlorinated compounds	PCBs, chlorinated dioxins	312
Aromatic amines	Methyl anilines such as *o*-toluidine	313
Various heavy metal salts and complexes	Lead, nickel and cadmium residues	314

Routine testing for potential mutagenic impurities can add to the overall cost of the process, and hence should be minimised if it is safe to do so. Current ICH guidelines [301] offer the option of dispensing with testing if process controls can be definitely shown to eliminate the impurity by purging (removal) in particular steps. Procedures for the calculation of purge factors for impurities have been published [318], and a recent survey of pharmaceutical manufacturers suggested that many companies were using such strategies to demonstrate control of mutagenic impurity levels [319].

Mutagenic and carcinogenic impurities can cause serious problems, so it is important to thoroughly consider whether any could be formed in any particular process and then test to ensure none are present. As is so often the case in process development, it is far better to anticipate problems at the planning stage rather than to wait for them to make their unwelcome presence felt at a later date.

Bibliography

[1]   Stork G, Kraus GA, Garcia GA. Regiospecific aldol condensations of the kinetic lithium enolates of methyl ketones. J Org Chem, 1974, 39, 3459–3460.
[2]   Groves JK. The Friedel-Crafts acylation of alkenes. Chem Soc Rev, 1972, 1, 73–97.
[3]   Razdan RK, Dalzell HC, Handrick GR. Hashish.[1] Simple one-step synthesis of (-)-Δ^1-tetrahydrocannabinol (THC) from p-metha-2,8-dien-1-ol and olivetol. J Am Chem Soc, 1974, 96, 5860–5865.
[4]   Aguillón AR, Leão RAC, De Oliveira KT, Brocksom TJ, Miranda LSM, De Souza ROMA. Process intensification for obtaining a cannabidiol intermediate by photo-oxygenation of limonene under continuous-flow conditions. Org Process Res Dev, 2020, 24, 2017–2024.
[5]   Fehr C, Ohloff G. US Patent 4433183 (Firmenich), 1984.
[6]   Burdick DC, Collier SJ, Jos F et al. US Patent 7674922 (Albany Molecular Research), 2010.
[7]   Sell CS. Ingredients for the modern perfumery industry. In: Pybus DH, Sell CS, eds. The chemistry of fragrances: From perfumer to consumer. 2nd ed, Cambridge, UK, Royal Society of Chemistry, 2006, 88–91.
[8]   Yang S, Tian H, Sun B, Liu Y, Hao Y, Lv Y. One-pot synthesis of (-)-Ambrox. Sci Rep, 2016, 6, 32650.
[9]   Hanessian S, Giroux S, Merner BL. Design and strategy in organic synthesis: From the chiron approach to catalysis. Weinheim, Germany, Wiley-VCH, 2013.
[10] Henschke JP, Liu Y, Chen Y-F, Meng D, Sun T. US Patent 7897795 (Scinopharm), 2011.
[11] St Jernschantz JW, Resul B. European Patent 364417 B1 (Pharmacia), 1994.
[12] Resul B. US Patent 5359095 (Pharmacia), 1994.
[13] Greenwood AK, McHattie D, Thompson DG, Clissold DW. World Patent Application WO 02/096898 (Resolution Chemicals and Cascade Biochem), 2002.
[14] Boulton LT, Brick D, Fox ME et al. Synthesis of the potent antiglaucoma agent, travoprost. Org Process Res Dev, 2002, 6, 138–145.
[15] Boyer T, Choudary BM, Edwards AJ et al. Development of a scalable process for the PPAR-α agonist GW641597X incorporating Baeyer-Villiger chemistry and retrospective ICH M7 assessment. Org Process Res Dev, 2020, 24, 371–386.
[16] Wenzl P, Dürrholz F, Diehl HV. US Patent 7205440 (Lanxess), 2007.
[17] Pfirmann R. US Patent 5410082 (Hoechst), 1995.
[18] Sell CS. Ingredients for the modern perfumery industry. In: Pybus DH, Sell CS, eds. The chemistry of fragrances: From perfumer to consumer. 2nd ed, Cambridge, UK, Royal Society of Chemistry, 2006, 61.
[19] Klucznik T, Mikulak-Klucznik B, McCormack MP et al. Efficient syntheses of diverse, medicinally relevant targets planned by computer and executed in the laboratory. Chem, 2018, 4, 522–532.
[20] Badowski T, Molga K, Grzybowski BA. Selection of cost-effective yet chemically diverse pathways from the networks of computer-generated retrosynthetic plans. Chem Sci, 2019, 10, 4640–4651.
[21] Lu J, Toy PH. Organic polymer supports for synthesis and for reagent and catalyst immobilization. Chem Rev, 2009, 109, 815–838.
[22] Nagendrappa G. Organic synthesis using clay and clay-supported catalysts. Appl Clay Sci, 2011, 53, 106–138.
[23] Bhattacharyya S, Basu B. Solid-supported catalysis. In: Zhang W, Cue BW, eds. Green techniques for organic synthesis and medicinal chemistry. 2nd ed, Hoboken, NJ, USA, John Wiley & Sons, 2018, 269–289.

https://doi.org/10.1515/9783110717877-009

[24] Noyori R, Takeshi O. Asymmetric catalysis by architectural and functional molecular engineering: Practical chemo- and stereoselective hydrogenation of ketones. Angew Chem Int Ed, 2001, 40, 40–73.

[25] Guerreiro P, Ratovelomanana-Vidal V, Genêt J-P, Dellis P. Recyclable diguanidinium-BINAP and PEG-BINAP supported catalysts: Syntheses and use in Rh(I) and Ru(II) asymmetric hydrogenation reactions. Tetrahedron Lett, 2001, 42, 3423–3426.

[26] McDonald AR, Müller C, Vogt D, Van Klink GPM, Van Koten G. BINAP-Ru and -Rh catalysts covalently immobilised on silica and their repeated application in asymmetric hydrogenation. Green Chem, 2008, 10, 424–432.

[27] Castro LCM, Li H, Sortais J-B, Darcel C. When iron met phosphines: A happy marriage for reduction catalysis. Green Chem, 2015, 17, 2283–2303.

[28] Wei D, Darcel C. Iron catalysis in reduction and hydrometalation reactions. Chem Rev, 2019, 119, 2550–2610.

[29] Kim J, Suri JT, Cordes DB, Singaram B. Asymmetric reductions involving borohydrides: A practical asymmetric reduction of ketones mediated by (L)-TarB–NO_2: A chiral Lewis acid. Org Process Res Dev, 2006, 10, 949–958.

[30] Ren Y, Tian X, Sun K, Xu J, Xu X, Lu S. Highly enantioselective reduction of ketones by chiral diol-modified lithium aluminum hydride reagents. Tetrahedron Lett, 2006, 47, 463–465.

[31] Corey EJ, Helal CJ. Reduction of carbonyl compounds with chiral oxazaborolidine catalysts: A new paradigm for enantioselective catalysis and a powerful new synthetic method. Angew Chem Int Ed, 1998, 37, 1986–2012.

[32] Masui M, Shioiri T. A practical method for asymmetric borane reduction of prochiral ketones using chiral amino alcohols and trimethyl borate. Synlett, 1997, 273–274.

[33] Kawanami H, Yanagita RC. Practical enantioselective reduction of ketones using oxazaborolidine catalysts generated in situ from chiral lactam alcohols. Molecules, 2018, 23, 2408.

[34] Lidder S, Dargan PI, Sexton M et al. Cardiovascular toxicity associated with recreational use of diphenylprolinol (diphenyl-2-pyrrolidinemethanol [D2PM]). J Med Toxicol, 2008, 4, 167–169.

[35] Wang H-Y, Tang J-W, Peng P, Yan H-J, Zhang F-L, Chen S-X. Development of a novel chemoenzymatic process for (S)-1-(pyridin-4-yl)-1,3-propanediol. Org Process Res Dev, 2020, 24, 2890–2897.

[36] Huisman GW, Liang G, Krebber A. Practical chiral alcohol manufacture using ketoreductases. Curr Opin Chem Biol, 2010, 14, 122–129.

[37] Wolfson A, Dlugy C, Tavor D. Baker's yeast catalyzed asymmetric reduction of prochiral ketones in different reaction mediums. Org Commun, 2013, 6, 1–11.

[38] Chen Q, Wang K, Yuan C. A chemo-enzymatic synthesis of chiral secondary alcohols bearing sulfur-containing functionality. New J Chem, 2009, 33, 972–975.

[39] Rodríguez S, Kayser M, Stewart JD. Improving the stereoselectivity of bakers' yeast reductions by genetic engineering. Org Lett, 1999, 1, 1153–1155.

[40] Yang Y, Drolet M, Kayser MM. The dynamic kinetic resolution of 3-oxo-4-phenyl-β-lactam by recombinant E. coli overexpressing yeast reductase Ara1p. Tetrahedron: Asymmetry, 2005, 16, 2748–2753.

[41] Wei T-Y, Tang J-W, Ni G-W et al. Development of an enzymatic process for the synthesis of (S)-2-chloro-1-(2,4-dichlorophenyl)ethanol. Org Process Res Dev, 2019, 23, 1822–1828.

[42] The Merck Index. 15th ed, Cambridge, UK, Royal Society of Chemistry, 2013, 704.

[43] Yang Q, Sheng M, Huang Y. Potential safety hazards associated with using N,N-dimethylformamide in chemical reactions. Org Process Res Dev, 2020, 24, 1586–1601.

[44] Wilkinson MC. "Greener" Friedel–Crafts acylations: A metal- and halogen-free methodology. Org Lett, 2011, 13, 2232–2235.

[45] Zhang B, Zou N, Wang W, Wang Z. Investigation of an accidental explosion in a nitromethane rectification process. Process Saf Environ Prot, 2015, 94, 358–365.

[46] Martínez MA, Ballesteros S, Almarza E, Sánchez De La Torre C, Búa S. Acute nitrobenzene poisoning with severe associated methemoglobinemia: Identification in whole blood by GC-FID and GC-MS. J Anal Toxicol, 2003, 27, 221–225.

[47] Hallett JP, Welton T. Room-temperature ionic liquids: Solvents for synthesis and catalysis. 2. Chem Rev, 2011, 111, 3508–3576.

[48] Pedro SN, Freire CSR, Silvestre AJD, Freire MG. The role of ionic liquids in the pharmaceutical field: An overview of relevant applications. Int J Mol Sci, 2020, 21, 8298.

[49] Laird T. Cost estimates for new molecules. Org Process Res Dev, 2005, 9, 125.

[50] Carr-Brion KG. Personal communication.

[51] Craig DQM. Pharmaceutical applications of DSC. In: Craig DQM, Reading M, eds. Thermal analysis of pharmaceuticals. Boca Raton, FL, USA, CRC Press, 2006, 53–100.

[52] Hada S, Harrison BK. Adiabatic-temperature rise: An awkward calculation made simple. Chem Eng, 2007, 114, 45–48.

[53] Allian AD, Shah NP, Ferretti AC, Brown DB, Kolis SP, Sperry JB. Process safety in the pharmaceutical industry – Part 1: Thermal and reaction hazard evaluation processes and techniques. Org Process Res Dev, 2020, 24, 2529–2548.

[54] Roduit B, Hartmann M, Folly P, Sarbach A, Brodard P, Baltensperger R. Thermal decomposition of AIBN, Part B: Simulation of SADT value based on DSC results and large scale tests according to conventional and new kinetic merging approach. Thermochim Acta, 2015, 621, 6–24.

[55] Greenwood AK, McHattie D, Bhatarah P. European Patent 1608670 B1 (Resolution Chemicals Ltd), 2013.

[56] Peters BK, Rodriguez KX, Reisberg SH et al. Scalable and safe synthetic organic electroreduction inspired by Li-ion battery chemistry. Science, 2019, 363, 838–845.

[57] Monteiro AM, Flanagan RC. Process safety considerations for the use of 1M borane tetrahydrofuran complex under general purpose plant conditions. Org Process Res Dev, 2017, 21, 241–246.

[58] Stoessel F. Experimental study of thermal hazards during the hydrogenation of aromatic nitro compounds. J Loss Prev Process Ind, 1993, 6, 79–85.

[59] Nobis M, Roberge DM. Mastering ozonolysis: Production from laboratory to ton scale in continuous flow. Chim Oggi, 2011, 29, 56–58.

[60] Sheng M, Frurip D, Gorman D. Reactive chemical hazards of diazonium salts. J Loss Prev Process Ind, 2015, 38, 114–118.

[61] Sullivan JM. Explosion during preparation of benzenediazonium-2-carboxylate hydrochloride. J Chem Educ, 1971, 48, 419.

[62] Firth JD, Fairlamb IJS. A need for caution in the preparation and application of synthetically versatile aryl diazonium tetrafluoroborate salts. Org Lett, 2020, 22, 7057–7059.

[63] Ullrich R, Grewer T. Decomposition of aromatic diazonium compounds. Thermochim Acta, 1993, 225, 201–211.

[64] Oger N, Le Grognec E, Felpin F-X. Handling diazonium salts in flow for organic and material chemistry. Org Chem Front, 2015, 2, 590–614.

[65] Shukla CA, Kulkarni AA, Ranade VV. Selectivity engineering of the diazotization reaction in a continuous flow reactor. React Chem Eng, 2016, 1, 387–396.

[66] Teci M, Tilley M, McGuire MA, Organ MG. Handling hazards using continuous flow chemistry: Synthesis of N^1-aryl-[1,2,3]-triazoles from anilines via telescoped three-step diazotization, azidodediazotization, and [3 + 2] dipolar cycloaddition processes. Org Process Res Dev, 2016, 20, 1967–1973.

[67] Health & Safety Executive. The Fire at Hickson & Welch Limited. Sudbury, UK, HSE Books, 1994.

[68] Yang Q, Sheng M, Li X et al. Potential explosion hazards associated with the autocatalytic thermal decomposition of dimethyl sulfoxide and its mixtures. Org Process Res Dev, 2020, 24, 916–939.

[69] Deguchi Y, Kono M, Koizumi Y, Izato Y-I, Miyake A. Study on autocatalytic decomposition of dimethyl sulfoxide (DMSO). Org Process Res Dev, 2020, 24, 1614–1620.

[70] Yang Q, Sheng M, Henkelis JJ et al. Explosion hazards of sodium hydride in dimethyl sulfoxide, N,N-dimethylformamide, and N,N-dimethylacetamide. Org Process Res Dev, 2019, 23, 2210–2217.

[71] Wang Z, Richter SM, Rozema MJ, Schellinger A, Smith K, Napolitano JG. Potential safety hazards associated with using acetonitrile and a strong aqueous base. Org Process Res Dev, 2017, 21, 1501–1508.

[72] Rivera NR, Kassim B, Grigorov P et al. Investigation of a flow step clogging incident: A precautionary note on the use of THF in commercial-scale continuous process. Org Process Res Dev, 2019, 23, 2556–2561.

[73] Urben PG. Bretherick's handbook of reactive chemical hazards. 8th ed, Amsterdam, Netherlands, Elsevier, 2017.

[74] Stoessel F. Thermal safety of chemical processes: Risk assessment and process design. 2nd ed, Weinheim, Germany, Wiley-VCH, 2020.

[75] Dixon Jr F, Flanagan RC. Reductive amination bicarbonate quench: Gas-evolving waste stream near-miss investigation. Org Process Res Dev, 2020, 24, 1063–1067.

[76] Cotarca L, Geller T, Répási J. Bis(trichloromethyl)carbonate (BTC, triphosgene): A safer alternative to phosgene?. Org Process Res Dev, 2017, 21, 1439–1446.

[77] Kurita K, Iwakura Y. Trichloromethyl chloroformate as a phosgene equivalent: 3-Isocyanatopropanoyl chloride. Org Synth, 1979, 59, 195.

[78] Nowakowski J. Isocyanate intermediates. II. Trichloromethyl chloroformate – A convenient reagent for the preparation of diisocyanates with benzene or furan rings. J Prakt Chem, 1992, 334, 187–189.

[79] Rannard SP, Davis NJ. Controlled synthesis of asymmetric dialkyl and cyclic carbonates using the highly selective reactions of imidazole carboxylic esters. Org Lett, 1999, 1, 933–936.

[80] Rivetti F. The role of dimethylcarbonate in the replacement of hazardous chemicals. C R Acad Sci Sér IIC, 2000, 3, 497–503.

[81] Ion A, Van Doorslaer C, Parvulescu V, Jacobs P, De Vos D. Green synthesis of carbamates from CO_2, amines and alcohols. Green Chem, 2008, 10, 111–116.

[82] Sharma SK, Agarwal DD. Oxidative chlorination of aromatic compounds in aqueous media. J Agric Life Sci, 2014, 1, 146–164.

[83] Xin H, Yang S, An B, An Z. Selective water-based oxychlorination of phenol with hydrogen peroxide catalyzed by manganous sulfate. RSC Adv, 2017, 7, 13467–13472.

[84] Steiner A, Roth PMC, Strauss FJ et al. Multikilogram per hour continuous photochemical benzylic brominations applying a smart dimensioning scale-up strategy. Org Process Res Dev, 2020, 24, 2208–2216.

[85] Park SB. Alert over South Korea toxic leaks. Nature, 2013, 494, 15–16.

[86] Muriale L, Lee E, Genovese J, Trend S. Fatality due to acute fluoride poisoning following dermal contact with hydrofluoric acid in a palynology laboratory. Ann Occup Hyg, 1996, 40, 705–710.

[87] Lam KK, Lau FL. An incident of hydrogen cyanide poisoning. Am J Emerg Med, 2000, 18, 172–175.

[88] Cyanide poisoning – Recommendations on first aid treatment for employers and first aiders. National poisons information service. Birmingham, UK. (Accessed January 26th, 2021, at https://www.npis.org/Download/Cyanide%20Guidance.pdf).

[89] Memoli S, Selva M, Tundo P. Dimethylcarbonate for eco-friendly methylation reactions. Chemosphere, 2001, 43, 115–121.

[90] Yan H, Zeng L, Xie Y, Cui Y, Ye L, Tu S. N-Methylation of poorly nucleophilic aromatic amines with dimethyl carbonate. Res Chem Intermed, 2016, 42, 5951–5960.

[91] Selva M, Perosa A. Green chemistry metrics: A comparative evaluation of dimethyl carbonate, methyl iodide, dimethyl sulfate and methanol as methylating agents. Green Chem, 2008, 10, 457–464.

[92] Fuhrmann E, Talbiersky J. Synthesis of alkyl aryl ethers by catalytic Williamson ether synthesis with weak alkylation agents. Org Process Res Dev, 2005, 9, 206–211.

[93] Afanasyev OI, Kuchuk E, Usanov DL, Chusov D. Reductive amination in the synthesis of pharmaceuticals. Chem Rev, 2019, 119, 11857–11911.

[94] Bataille CJR, Donohoe TJ. Osmium-free direct *syn*-dihydroxylation of alkenes. Chem Soc Rev, 2011, 40, 114–128.

[95] Horiuchi CA, Dan G, Sakamoto M et al. A new synthesis of *cis*-diol from alkene using iodine-ammonium cerium(IV) nitrate. Synthesis, 2005. 2861–2864.

[96] Bielski R, Joyce PJ. The use of methyltricaprylylammonium chloride as a phase transfer catalyst for the destruction of methyl bromide in air streams. Org Process Res Dev, 2008, 12, 781–784.

[97] Elsariti SM. Haftirman. Behaviour of stress corrosion cracking of austenitic stainless steels in sodium chloride solutions. Procedia Eng, 2013, 53, 650–654.

[98] Douglas JJ, Adams BWV, Benson H et al. Multikilogram-scale preparation of AZD4635 via C–H borylation and bromination: The corrosion of tantalum by a bromine/methanol mixture. Org Process Res Dev, 2019, 23, 62–68.

[99] Crawley F, Tyler B. HAZOP: Guide to best practice. 3rd ed, Amsterdam, Netherlands, Elsevier, 2015.

[100] Rocheleau M-J. Analytical methods for determination of counter-ions in pharmaceutical salts. Curr Pharm Anal, 2008, 4, 25–32.

[101] USF-NF. US Pharmacopeia. (Accessed February 4th 2021 at https://www.uspnf.com).

[102] European Pharmacopoeia. 10th ed, Strasbourg, France, European Directorate for the Quality of Medicines and Healthcare, 2019.

[103] Yu LX, Amidon G, Khan MA et al. Understanding pharmaceutical quality by design. AAPS J, 2014, 16, 771–783.

[104] Mojica CA, St Pierre-Berry L, Sistare F. Process analytical technology in the manufacture of bulk active pharmaceuticals – Promise, practice, and challenges. In: Gadamasetti K, Braish T, eds. Process chemistry in the pharmaceutical industry. Vol. 2, Boca Raton, FL, USA, CRC Press, 2008, 361–382.

[105] Mitic A, Cervera-Padrell AE, Mortensen AR et al. Implementation of near-infrared spectroscopy for in-line monitoring of a dehydration reaction in a tubular laminar reactor. Org Process Res Dev, 2016, 20, 395–402.

[106] Lopes LC, Brandão IV, Sánchez OC et al. Horseradish peroxidase biocatalytic reaction monitoring using near-infrared (NIR) spectroscopy. Process Biochem, 2018, 71, 127–133.

[107] Otsuka M. Near-infrared spectroscopy application to the pharmaceutical industry. In: Meyers RA, ed. Encyclopedia of analytical chemistry. New York, USA, Wiley, 2000.

[108] Hamminga GM, Mul G, Moulijn GA. Applicability of fiber-optic-based Raman probes for on-line reaction monitoring of high-pressure catalytic hydrogenation reactions. Appl Spectrosc, 2007, 61, 470–478.

[109] Kadić A, Chylenski P, Hansen MAT, Bengtsson O, Eijsink VGH, Lidén G. Oxidation-reduction potential (ORP) as a tool for process monitoring of H_2O_2/LPMO assisted enzymatic hydrolysis of cellulose. Process Biochem, 2019, 86, 89–97.

[110] Ahmed-Omer B, Sliwinski E, Cerroti JP, Ley SV. Continuous processing and efficient *in situ* reaction monitoring of a hypervalent iodine(III) mediated cyclopropanation using benchtop NMR spectroscopy. Org Process Res Dev, 2016, 20, 1603–1614.

[111] Elipe MVS, Cherney A, Krull R et al. Application of the new 400 MHz high-temperature superconducting (HTS) power-driven magnet NMR technology for online reaction monitoring: Proof of concept with a ring-closing metathesis (RCM) reaction. Org Process Res Dev, 2020, 24, 1428–1434.

[112] Sun J, Yin Y, Li W, Jin O, Na N. Chemical reaction monitoring by ambient mass spectrometry. Mass Spectrom Rev, 2022, 41, 70–99.

[113] Ma G, Jha A. A practical guide for buffer-assisted isolation and purification of primary, secondary, and tertiary amine derivatives from their mixture. Org Process Res Dev, 2005, 9, 847–852.

[114] Boucher MM, Furigay MH, Quach PK, Brindle CS. Liquid–liquid extraction protocol for the removal of aldehydes and highly reactive ketones from mixtures. Org Process Res Dev, 2017, 21, 1394–1403.

[115] Kister HZ. Distillation Design. New York, USA, McGraw Hill, 1992.

[116] Royals EE. Cyclization of pseudoionone by acidic reagents. Ind Eng Chem, 1946, 38, 546–548.

[117] Hertel O, Kiefer H, Arnold L. US Patent 4565894 (BASF), 1986.

[118] Seitz K, Günthard HH, Jeger O. Veilchenriechstoffe. 37. Mitteilung. Über die Trennung von α- und β-Jonon durch fraktionierte Destillation. Helv Chim Acta, 1950, 33, 2196–2202.

[119] Shu M-J, Pan H-L, Wu T-G, Dai W-L. Separation of ionone isomers by vacuum batch distillation. Xiandai Huagong/Mod Chem Ind, 2017, 37, 160–163.

[120] Lorenz H-M, Staak D, Grützner T, Repke J-U. Divided wall columns: Usefulness and challenges. Chem Eng Trans, 2018, 69, 229–234.

[121] Papakadis E, Tula AK, Gani R. Solvent selection methodology for pharmaceutical processes: Solvent swap. Chem Eng Res Des, 2016, 115, 443–461.

[122] Urwin SJ, Levilain G, Marziano I, Merritt JM, Houson I, Ter Horst JH. A structured approach to cope with impurities during industrial crystallization development. Org Process Res Dev, 2020, 24, 1443–1456.

[123] Leyssens T, Baudry C, Hernandez MLE. Optimization of a crystallization by online FBRM analysis of needle-shaped crystals. Org Process Res Dev, 2011, 15, 413–426.

[124] Acevedo D, Wu W-L, Yang X, Pavurala N, Mohammad A, O'Connor TF. Evaluation of focused beam reflectance measurement (FBRM) for monitoring and predicting the crystal size of carbamazepine in crystallization processes. CrystEngComm, 2021, 23, 972–985.

[125] Schaefer C, Lecomte C, Clicq D, Merschaert A, Norrant E, Fotiadu F. On-line near infrared spectroscopy as a process analytical technology (PAT) tool to control an industrial seeded API crystallization. J Pharm Biomed Anal, 2013, 83, 194–201.

[126] Wood B, Girard KP, Polster CS, Croker DM. Progress to date in the design and operation of continuous crystallization processes for pharmaceutical applications. Org Process Res Dev, 2019, 23, 122–144.

[127] Orehek J, Teslić D, Likozar B. Continuous crystallization processes in pharmaceutical manufacturing: A review. Org Process Res Dev, 2021, 25, 16–42.

[128] Ma Y, Wu S, Macaringue EGJ, Zhang T, Gong J, Wang J. Recent progress in continuous crystallization of pharmaceutical products: Precise preparation and control. Org Process Res Dev, 2020, 24, 1785–1801.

[129] Benitez-Chapa AG, Nigam KDP, Alvarez AJ. Process intensification of continuous antisolvent crystallization using a coiled flow inverter. Ind Eng Chem Res, 2020, 59, 3934–3942.

[130] Nohira H, Sakai K. Optical resolution by means of crystallization. In: Toda F, ed. Enantiomer separation: Fundamentals and practical methods. Dordrecht, The Netherlands, Kluwer Academic Publishers, 2004, 165–191.

[131] Wang Y, Chen A. Crystallization-based separation of enantiomers. In: Andrushko V, Andrushko N, eds. Stereoselective synthesis of drugs and natural products. Hoboken, NJ, USA, John Wiley & Sons, 2013, 1663–1682.

[132] Aelterman W, Lang Y, Willemsens B, Vervest I, Leurs S, De Knaep F. Conversion of the laboratory synthetic route of the N-aryl-2-benzothiazolamine R116010 to a manufacturing method. Org Process Res Dev, 2001, 5, 467–471.

[133] Simon M, Donnellan P, Glennon B, Jones RC. Resolution via diastereomeric salt crystallization of ibuprofen lysine: Ternary phase diagram studies. Chem Eng Technol, 2018, 41, 921–927.

[134] Stavber G, Cluzeau J. World Patent Application WO 2015/007897 (Lek Pharmaceuticals), 2015.

[135] Mossotti M, Barozza A, Roletto J, Paissoni P. World Patent Application WO 2017/072216 (Procos), 2017.

[136] Den Brok MWJ, Nuijen B, Lutz C, Opitz H-G, Beijnen JH. Pharmaceutical development of a lyophilised dosage form for the investigational anticancer agent imexon using dimethyl sulfoxide as solubilising and stabilising agent. J Pharm Sci, 2005, 94, 1101–1114.

[137] Ziaee A, Albadarin AB, Padrela L, Femmer T, O'Reilly E, Walker G. Spray drying of pharmaceuticals and biopharmaceuticals: Critical parameters and experimental process optimization approaches. Eur J Pharm Sci, 2019, 127, 300–318.

[138] Moussa Z, Judeh ZMA, Ahmed SA. Polymer-supported triphenylphosphine: Application in organic synthesis and organometallic reactions. RSC Adv, 2019, 9, 35217–35272.

[139] Thompson SK, Heathcock CH. Effect of cation, temperature, and solvent on the stereoselectivity of the Horner-Emmons reaction of trimethyl phosphonoacetate with aldehydes. J Org Chem, 1990, 55, 3386–3388.

[140] Beddoe RH, Andrews KG, Magné V et al. Redox-neutral organocatalytic Mitsunobu reactions. Science, 2019, 365, 910–914.

[141] Valeur E, Bradley M. Amide bond formation: Beyond the myth of coupling reagents. Chem Soc Rev, 2009, 38, 606–631.

[142] Miyamoto H, Sakumoto C, Takekoshi E et al. Effective method to remove metal elements from pharmaceutical intermediates with polychelated resin scavenger. Org Process Res Dev, 2015, 19, 1054–1061.

[143] Ren H, Strulson CA, Humphrey G et al. Potassium isopropyl xanthate (PIX): An ultra-efficient palladium scavenger. Green Chem, 2017, 19, 4002–4006.

[144] Mori K. Chemical synthesis of hormones, pheromones and other bioregulators. Chichester, West Sussex, UK, John Wiley & Sons, 2010.

[145] Xu J, Luo J, Kong L. Single-step preparative extraction of artemisinin from *Artemisia annua* by charcoal column chromatography. Chromatographia, 2011, 74, 471–475.

[146] Gerberding SJ, Byers CH. Preparative ion-exchange chromatography of proteins from dairy whey. J Chromatogr A, 1998, 808, 141–151.

[147] Edwards J. Large-scale column chromatography: A GMP manufacturing perspective. In: Goldberg E, ed. Handbook of downstream processing. London, UK, Chapman & Hall, 1997, 167–184.

[148] Dapremont O. Use of industrial scale chromatography in pharmaceutical manufacturing. In: Knochel P, Molander GA, eds. Comprehensive organic synthesis. 2nd ed. Amsterdam, The Netherlands, Elsevier, Vol. 9, 2014, 181–206.

[149] Gomes PS, Rodrigues AE. Simulated moving bed chromatography: From concept to proof-of-concept. Chem Eng Technol, 2012, 35, 17–34.

[150] Kim K-M, Lee JW, Kim S, Da Silva FVS, Seidel-Morgenstern A, Lee C-H. Advanced operating strategies to extend the applications of simulated moving bed chromatography. Chem Eng Technol, 2017, 40, 2163–2178.

[151] Rajendran A, Paredes G, Mazzotti M. Simulated moving bed chromatography for the separation of enantiomers. J Chromatogr A, 2009, 1216, 709–738.

[152] Pais LS, Mata VG, Rodrigues AE. Simulated moving bed and related techniques. In: Cox GB, ed. Preparative enantioselective chromatography. Oxford, UK, Blackwell Publishing, 2005, 176–198.

[153] Pflum DA, Wilkinson HS, Tanoury GJ et al. A large-scale synthesis of enantiomerically pure cetirizine dihydrochloride using preparative chiral HPLC. Org Process Res Dev, 2001, 5, 110–115.

[154] Speybrouck D, Lipka E. Preparative supercritical fluid chromatography: A powerful tool for chiral separations. J Chromatogr A, 2016, 1467, 33–55.

[155] Martin AJP, Synge RLM. A new form of chromatogram employing two liquid phases. Biochem J, 1941, 35, 1358–1368.

[156] Knox JH, Pyper HM. Framework for maximizing throughput in preparative liquid chromatography. J Chromatogr A, 1986, 363, 1–30.

[157] Samanidou VF. Basic LC method development and optimization. In: Anderson JL, Berthod A, Pino V, Stalcup AM, eds. Analytical separation science. Weinheim, Germany, Wiley-VCH, Vol. 1, 2015, 25–42.

[158] Van Deemter JJ, Zuiderweg FJ, Klinkenberg A. Longitudinal diffusion and resistance to mass transfer as causes of nonideality in chromatography. Chem Eng Sci, 1956, 5, 271–289.

[159] Gritti F, Guiochon G. The van Deemter equation: Assumptions, limits, and adjustment to modern high performance liquid chromatography. J Chromatogr A, 2013, 1302, 1–13.

[160] Aguiar AJ, Zelmer JE. Dissolution behavior of polymorphs of chloramphenicol palmitate and mefenamic acid. J Pharm Sci, 1969, 58, 983–987.

[161] Atici EB, Karliğa B. Quantitative determination of two polymorphic forms of imatinib mesylate in a drug substance and tablet formulation by X-ray powder diffraction, differential scanning calorimetry and attenuated total reflectance Fourier transform infrared spectroscopy. J Pharm Biomed Anal, 2015, 114, 330–340.

[162] Donahue M, Botonjic-Sehic E, Wells D, Brown CW. Understanding infrared and Raman spectra of pharmaceutical polymorphs. American Pharm Rev, 2011, 14, 104–110.

[163] Barrio M, Huguet J, Rietveld IB, Robert B, Céolin R, Tamarit J-L. The pressure-temperature phase diagram of metacetamol and its comparison to the phase diagram of paracetamol. J Pharm Sci, 2017, 106, 1538–1544.

[164] Telford R, Seaton CC, Clout A et al. Stabilisation of metastable polymorphs: The case of paracetamol form III. Chem Commun, 2016, 52, 12028–12031.

[165] Dunitz JD, Benstein J. Disappearing polymorphs. Acc Chem Res, 1995, 28, 193–200.

[166] Bučar D-K, Lancaster RW, Bernstein J. Disappearing polymorphs revisited. Angew Chem Int Ed, 2015, 54, 6972–6993.

[167] Morissette SL, Almarsson O, Peterson ML et al. High-throughput crystallization: Polymorphs, salts, co-crystals and solvates of pharmaceutical solids. Adv Drug Del Rev, 2004, 56, 275–300.

[168] Parambil JV, Poornachary SK, Heng JYY, Tan RBH. Template-induced nucleation for controlling crystal polymorphism: From molecular mechanisms to applications in pharmaceutical processing. CrystEngComm, 2019, 21, 4122–4135.

[169] Rengarajan GT, Enke D, Steinhart M, Beiner M. Size-dependent growth of polymorphs in nanopores and Ostwald's step rule of stages. Phys Chem Chem Phys, 2011, 13, 21367–21374.

[170] Diao Y, Whaley KE, Helgeson MR et al. Gel-induced selective crystallization of polymorphs. J Am Chem Soc, 2012, 134, 673–684.

[171] Saikia B, Mulvee MT, Torres-Moya I, Sarma B, Steed JW. Drug mimetic organogelators for the control of concomitant crystallization of barbital and thalidomide. Cryst Growth Des, 2020, 20, 7989–7996.

[172] Guerain M. A review on high pressure experiments for study of crystallographic behavior and polymorphism of pharmaceutical materials. J Pharm Sci, 2020, 109, 2640–2653.

[173] Fabbiani FBA, Allan DR, David WIF et al. High-pressure studies of pharmaceuticals: An exploration of the behavior of piracetam. Cryst Growth Des, 2007, 7, 1115–1124.

[174] Zakharov BA, Seryotkin YV, Tumanov NA et al. The role of fluids in high-pressure polymorphism of drugs: Different behaviour of β-chlorpropamide in different inert gas and liquid media. RSC Adv, 2016, 6, 92629–92637.

[175] Belenguer AM, Lampronti GI, Cruz-Cabeza AJ, Hunter CA, Sanders JKM. Solvation and surface effects on polymorph stabilities at the nanoscale. Chem Sci, 2016, 7, 6617–6627.

[176] Solomos MA, Capacci-Daniel C, Rubinson JF, Swift JA. Polymorph selection via sublimation onto siloxane templates. Cryst Growth Des, 2018, 18, 6965–6972.

[177] Kamali N, O'Malley C, Mahon MF, Erxleben A, McArdle P. Use of sublimation catalysis and polycrystalline powder templates for polymorph control of gas phase crystallization. Cryst Growth Des, 2018, 18, 3510–3516.

[178] Liu Z, Li C. Solvent-free crystallizations of amino acids: The effects of the hydrophilicity/hydrophobicity of side-chains. Biophys Chem, 2008, 138, 115–119.

[179] Bernstein J, MacAlpine J. Pharmaceutical crystal forms and crystal-form patents: Novelty and obviousness. In: Hilfiker R, Von Raumer M, eds. Polymorphism in the pharmaceutical industry: Solid form and drug development. Weinheim, Germany, Wiley-VCH, 2019, 469–483.

[180] Barker KV Polymorphs at the EPO: Where are we now? Finnegan European IP Blog, 2020 (Accessed March 11th, 2021, at https://www.finnegan.com/en/insights/blogs/european-ip-blog/polymorphs-at-the-epo-where-are-we-now.html).

[181] Tandon R, Tandon N, Thapar RK. Patenting of polymorphs. Pharm Pat Anal, 2018, 7, 59–63.

[182] Wieser G, Griesser U, Enders M, Kahlenberg V. European Patent 2726462 B1 (Sandoz), 2017.

[183] Bauer J, Spanton S, Henry R et al. Ritonavir: An extraordinary example of conformational polymorphism. Pharm Res, 2001, 18, 859–866.

[184] Chemburkar SR, Bauer J, Deming K et al. Dealing with the impact of ritonavir polymorphs on the late stages of bulk drug process development. Org Process Res Dev, 2000, 4, 413–417.

[185] Tawashi R. Aspirin: Dissolution rates of two polymorphic forms. Science, 1968, 160, 76.

[186] Vishweshwar P, McMahon JA, Oliveira M, Peterson ML, Zaworotko MJ. The predictably elusive form II of aspirin. J Am Chem Soc, 2005, 127, 16802–16803.

[187] Bond AD, Solanko KA, Parsons S, Redder S, Boese R. Single crystals of aspirin form II: crystallisation and stability. CrystEngComm, 2011, 13, 399–401.

[188] Crowell EL, Dreger ZA, Gupta YM. High-pressure polymorphism of acetylsalicylic acid (Aspirin): Raman spectroscopy. J Mol Struct, 2015, 1082, 29–37.

[189] Shtukenberg AG, Hu CT, Zhu Q et al. The third ambient aspirin polymorph. Cryst Growth Des, 2017, 17, 3562–3566.

[190] Yang J, Erriah B, Hu CT et al. A deltamethrin crystal polymorph for more effective malaria control. Proc Natl Acad Sci USA, 2020, 117, 26633–26638.

[191] Roth BD. US Patent 5273995 (Warner-Lambert), 1993.

[192] Briggs CA, Jennings RW, Wade R et al. US Patent 5969156 (Warner-Lambert), 1999.

[193] Shete G, Puri V, Kumar L, Bansal AK. Solid state characterization of commercial crystalline and amorphous atorvastatin calcium samples. AAPS PharmSciTech, 2010, 11, 598–609.

[194] Skorda D, Kontoyannis CG. Identification and quantitative determination of atorvastatin calcium polymorph in tablets using FT-Raman spectroscopy. Talanta, 2008, 74, 1066–1070.

[195] Roth BD. US Patent 4681893 (Warner-Lambert), 1987.

[196] McKenzie AT. European Patent 848704 B1 (Warner-Lambert), 2001.

[197] Mathew J, Ganesh S. World Patent Application WO 02/057229 (Biocon), 2002.

[198] Faustmann J, Jegorov A. World Patent Application WO 03/050085 (Ivax), 2003.

[199] Kumar Y, Thaper RK, Kumar SMD. US Patent 6528660 B1 (Ranbaxy), 2003.

[200] Byrn SB, Coates DA, Gushurst KS et al. US Patent 6605729 B1 (Warner-Lambert), 2003.

[201] Greff Z, Nagy PK, Barkoczy J et al. US Patent 6646133 B1 (Egis), 2003.

[202] Blatter F, Szelagiewicz M, Van Der Schaaf PA. WO 2004/050618 (Teva), 2004.

[203] Kumar Y, Kumar SMD, Sathyanarayana S. WO 2005/090301 (Ranbaxy), 2005.

[204] Grewal MS, Raj B, Singh J, Suri S. European Patent 1562583 A1 (Morepen Laboratories), 2005.

[205] Lidor-Hadas R, Niddam V, Lifshitz R, Kovalevski-Ishal E. US Patent 6992194 B2 (Teva), 2006.

[206] Krzyzaniak JF, Laurence Jr JM, Park A et al. World Patent Application WO 2006/011041 (Warner-Lambert), 2006.

[207] Lifshitz-Liron R, Aronhime J, Tessler L. World Patent Application WO 2006/012499 (Teva), 2006.

[208] Suri S, Sarin GS. World Patent Application WO 2006/048894 (Morepen Laboratories), 2006.

[209] Barkóczy J, Kótay Nagy P, Simig G et al. World Patent Application WO 2006/106372 (Egis), 2006.

[210] Manne SR, Chakilam N, Gudipati S, Katkam S, Sagyam RR. US Patent 7074818 B2 (Dr Reddy's), 2006.

[211] Gogulapati VPR, Chavakula R, Bandari M, Gorantla SR. World Patent Application WO 2007/096903 (Matrix Laboratories), 2007.

[212] Ayalon A, Levinger M, Roytblat S, Niddam V, Lifshitz R, Aronhime J. US Patent 7411075 B1 (Teva), 2008.

[213] An S-G, Sohn Y-T. Crystal forms of atorvastatin. Arch Pharm Res, 2009, 32, 933–936.

[214] Tessler L, Aronhime J, Lifshitz-Liron R, Maidan-Hanoch D, Hasson N. US Patent 7501450 B2 (Teva), 2009.

[215] Van Der Schaaf PA, Blatter F, Szelagiewicz M, Schöning K-U. US Patent 7538136 (Teva), 2009.

[216] Pinchasov M, Aronhime J, Doani Z. US Patent 8080672 B2 (Teva), 2011.

[217] Rao PRV, Somannaver YS, Kumar SN, Reddy SB, Islam A, Babu HB. Preparation of stable new polymorphic form of atorvastatin calcium. Der Pharm Lett, 2011, 3, 48–53.

[218] Chadha R, Kuhad A, Arora P, Kishor S. Characterisation and evaluation of pharmaceutical solvates of atorvastatin calcium by thermoanalytical and spectroscopic studies. Chem Cent J, 2012, 6, 114.

[219] Price SL. Predicting crystal structures of organic compounds. Chem Soc Rev, 2014, 43, 2098–2111.

[220] Price SL. Control and prediction of the organic solid state: A challenge to theory and experiment. Proc Royal Soc A, 2018, 474, 20180351.

[221] Habgood M. Form II caffeine: A case study for confirming and predicting disorder in organic crystals. Cryst Growth Des, 2011, 11, 3600–3608.

[222] Dichi E, Legendre B, Sghaier M. Physico-chemical characterisation of a new polymorph of caffeine. J Therm Anal Calorim, 2014, 115, 1551–1561.

[223] Klimeš J, Michaelides A. Perspective: Advances and challenges in treating van der Waals dispersion forces in density functional theory. J Chem Phys, 2012, 137, 120901.

[224] Burke K, Wagner LO. DFT in a nutshell. Int J Quantum Chem, 2013, 113, 96–101.

[225] Dybeck EC, Abraham NS, Schieber NP, Shirts MR. Capturing entropic contributions to temperature-mediated polymorphic transformations through molecular modeling. Cryst Growth Des, 2017, 17, 1775–1787.

[226] Reilly AM, Cooper RI, Adjiman CS et al. Report on the sixth blind test of organic crystal structure prediction methods. Acta Crystallogr B, 2016, B72, 439–459.

[227] Leardi R. Experimental design in chemistry: A tutorial. Anal Chim Acta, 2009, 652, 161–172.

[228] Uhlig N, Martins A, Gao D. Selective DIBAL-H monoreduction of a diester using continuous flow chemistry: From benchtop to kilo lab. Org Process Res Dev, 2020, 24, 2326–2335.

[229] Lamberto DJ, Neuhaus J. Robust process scale-up leveraging design of experiments to map active pharmaceutical ingredient humid drying parameter space. Org Process Res Dev, 2021, 25, 239–249.

[230] Cruz ACF, Mateus EM, Peterson MJ. Process development of a Sonogashira cross-coupling reaction as the key step of tirasemtiv synthesis using design of experiments. Org Process Res Dev, 2021, 25, 668–678.

[231] Montgomery DC. Design and analysis of experiments. 9th ed, Hoboken, NJ, USA, John Wiley & Sons, 2017.

[232] Caulcutt R. Statistics in Research and Development. 2nd ed, Boca Raton, FL, USA, CRC Press, 1991.

[233] Filipponi P, Ostacolo C, Novellino E, Pellicciari R, Gioiello A. Continuous flow synthesis of thieno[2,3-c]isoquinolin-5(4H)-one scaffold: A valuable source of PARP-1 inhibitors. Org Process Res Dev, 2014, 18, 1345–1353.

[234] Mennen SM, Alhambra C, Allen CL et al. The evolution of high-throughput experimentation in pharmaceutical development and perspectives on the future. Org Process Res Dev, 2019, 23, 1213–1242.

[235] Wickström G, Bendix T. The "Hawthorne effect" – What did the original Hawthorne studies actually show?. Scand J Work Environ Health, 2000, 26, 363–367.

[236] Box G, Kramer T. Statistical process monitoring and feedback adjustment: A discussion. Technometrics, 1992, 34, 251–267.

[237] Rosso V, Albrecht J, Roberts F, Janey JM. Uniting laboratory automation, DoE data, and modeling techniques to accelerate chemical process development. React Chem Eng, 2019, 4, 1646–1657.

[238] Nunn C, DiPietro A, Hodnett N, Sun P, Wells KM. High-throughput automated design of experiment (DOE) and kinetic modeling to aid in process development of an API. Org Process Res Dev, 2018, 22, 54–61.

[239] Akwi FM, Watts P. Continuous flow chemistry: Where are we now? Recent applications, challenges and limitations. Chem Commun, 2018, 54, 13894–13928.

[240] Shukla CA, Kulkarni AA. Automating multistep flow synthesis: Approach and challenges in integrating chemistry, machines and logic. Beilstein J Org Chem, 2017, 13, 960–987.

[241] Omoregbee K, Luc KNH, Dinh AH, Nguyen TV. Tropylium-promoted prenylation reactions of phenols in continuous flow. J Flow Chem, 2020, 10, 161–166.

[242] Wang Y-F, Jiang Z-H, Chu M-M et al. Asymmetric copper-catalyzed fluorination of cyclic β-keto esters in a continuous-flow microreactor. Org Biomol Chem, 2020, 18, 4927–4931.

[243] Kuleshova J, Hill-Cousins JT, Birkin PR, Brown RCD, Pletcher D, Underwood TJ. The methoxylation of N-formylpyrrolidine in a microfluidic electrolysis cell for routine synthesis. Electrochim Acta, 2012, 69, 197–202.

[244] Cova CM, Zuliani A, Manno R, Sebastian V, Luque R. Scrap waste automotive converters as efficient catalysts for the continuous-flow hydrogenations of biomass derived chemicals. Green Chem, 2020, 22, 1414–1423.

[245] Russo C, Gisbertz S, Williams JD et al. An oscillatory plug flow photoreactor facilitates semi-heterogeneous dual nickel/carbon nitride photocatalytic C–N couplings. React Chem Eng, 2020, 5, 597–604.

[246] Ouchi T, Battilocchio C, Hawkins JM, Ley SV. Process intensification for the continuous flow hydrogenation of ethyl nicotinate. Org Process Res Dev, 2014, 18, 1560–1566.

[247] Ingham RJ, Battilocchio C, Hawkins JM, Ley SV. Integration of enabling methods for the automated flow preparation of piperazine-2-carboxamide. Beilstein J Org Chem, 2014, 10, 641–652.

[248] Fitzpatrick DE, Battilocchio C, Ley SV. A novel internet-based reaction monitoring, control and autonomous self-optimization platform for chemical synthesis. Org Process Res Dev, 2016, 20, 386–394.

[249] Schaefer M, Stach E, Foitzik A. Computer controlled chemical micro-reactor. J Phys Conf Ser, 2006, 28, 024.

[250] Amara Z, Poliakoff M, Duque R et al. Enabling the scale-up of a key asymmetric hydrogenation step in the synthesis of an API using continuous flow solid-supported catalysis. Org Process Res Dev, 2016, 20, 1321–1327.

[251] King M. Process control: A practical approach. 2nd ed, Chichester, UK, Wiley, 2016.

[252] Karr CL, Sharma SK, Hatcher WJ, Harper TR. Fuzzy control of an exothermic chemical reaction using genetic algorithms. Eng Appl Artif Intell, 1993, 6, 575–582.

[253] Pandian BJ, Noel MM. Tracking control of a continuous stirred tank reactor using direct and tuned reinforcement learning based controllers. Chem Prod Process Mod, 2018, 13, 20170040.

[254] Shevlin M. Practical high-throughput experimentation for chemists. ACS Med Chem Lett, 2017, 8, 601–607.

[255] John A, Miranda MO, Ding K et al. Nickel catalysts for the dehydrative decarbonylation of carboxylic acids to alkenes. Organometallics, 2016, 35, 2391–2400.

[256] Welch CJ. High throughput analysis enables high throughput experimentation in pharmaceutical process research. React Chem Eng, 2019, 4, 1895–1911.

[257] Helmy R, Schafer W, Buhler L et al. Ambient pressure desorption ionization mass spectrometry in support of preclinical pharmaceutical development. Org Process Res Dev, 2010, 14, 386–392.

[258] Caputo M, Lyles JT, Salazar MS, Quave CL. Lego mindstorms fraction collector: A low-cost tool for a preparative high-performance liquid chromatography system. Anal Chem, 2020, 92, 1687–1690.

[259] Diaz D, De La Lglesia A, Barreto F, Borges R. DIY universal fraction collector. Anal Chem, 2021, 93, 9314–9318.

[260] Saitmacher K, Gabski H-P, Heider H, Patzlaff J, Wille C, Jung J. US Patent Application 2004/0131507A1 (Clariant), 2004.

[261] Mack AG. World Patent Application WO 96/11227 (Great Lakes), 1996.

[262] Sanderson K. Automation: Chemistry shoots for the moon. Nature, 2019, 568, 577–579.

[263] Angelone D, Hammer AJS, Rohrbach S et al. Convergence of multiple synthetic paradigms in a universally programmable chemical synthesis machine. Nat Chem, 2021, 13, 63–69.

[264] Stroo HF, Leeson A, Marqusee JA et al. Chlorinated ethene source remediation: Lessons learned. Environ Sci Technol, 2012, 46, 6438–6447.

[265] Jessop PG. Searching for green solvents. Green Chem, 2011, 13, 1391–1398.

[266] Krachko T, Lyaskovskyy V, Lutz M, Lammertsma K, Slootweg JC. Brønsted acid promoted reduction of tertiary phosphine oxides. Z Anorg Allg Chem, 2017, 643, 916–921.

[267] Manabe S, Wong CM, Sevov CS. Direct and scalable electroreduction of triphenylphosphine oxide to triphenylphosphine. J Am Chem Soc, 2020, 142, 3024–3031.

[268] Trost BM. The atom economy – A search for synthetic efficiency. Science, 1991, 254, 1471–1477.

[269] Kamata K, Ishimoto R, Hirano T, Kuzuya S, Uehara K, Mizuno N. Epoxidation of alkenes with hydrogen peroxide catalyzed by selenium-containing dinuclear peroxotungstate and kinetic, spectroscopic, and theoretical investigation of the mechanism. Inorg Chem, 2010, 49, 2471–2478.

[270] Sheldon RA. The E factor: Fifteen years on. Green Chem, 2007, 9, 1273–1283.

[271] Welton T. Solvents and sustainable chemistry. Proc R Soc A, 2015, 471, 20150502.

[272] Kleinekorte J, Fleitmann L, Bachmann M et al. Life cycle assessment for the design of chemical processes, products, and supply chains. Annu Rev Chem Biomol Eng, 2020, 11, 203–233.

[273] Jimenez-Gonzalez C, Ponder CS, Broxterman QB, Manley JB. Using the right green yardstick: Why process mass intensity is used in the pharmaceutical industry to drive more sustainable processes. Org Process Res Dev, 2011, 15, 912–917.

[274] Monteith ER, Mampuys P, Summerton L, Clark JH, Maes BUW, McElroy CR. Why we might be misusing process mass intensity (PMI) and a methodology to apply it effectively as a discovery level metric. Green Chem, 2020, 22, 123–135.

[275] World Meteorological Organization. State of the Global Climate 2020. Geneva, Switzerland, WMO, 2021.

[276] Belkhir L, Elmeligi A. Carbon footprint of the global pharmaceutical industry and relative impact of its major players. J Clean Prod, 2019, 214, 185–194.

[277] Liang B, He X, Hou J, Li L, Tang Z. Membrane separation in organic liquid: Technologies, achievements, and opportunities. Adv Mater, 2019, 31, 1806090.

[278] McGuinness EK, Zhang F, Ma Y, Lively RP, Losego MD. Vapor phase infiltration of metal oxides into nanoporous polymers for organic solvent separation membranes. Chem Mater, 2019, 31, 5509–5518.

[279] Crisenza GEM, Melchiorre P. Chemistry glows green with photoredox catalysis. Nat Commun, 2020, 11, 803.

[280] Laudadio G, Deng Y, Van Der Waal K et al. C(sp^3)–H functionalizations of light hydrocarbons using decatungstate photocatalysis in flow. Science, 2020, 369, 92–96.

[281] Kohli K, Prajapati R, Sharma BK. Bio-based chemicals from renewable biomass for integrated biorefineries. Energies, 2019, 12, 233.

[282] Dale BE. Feeding a sustainable chemical industry: Do we have the bioproducts cart before the feedstocks horse?. Faraday Discuss, 2017, 202, 11–30.

[283] Galanie S, Thodey K, Trenchard IJ, Interrante MF, Smolke CD. Complete biosynthesis of opioids in yeast. Nature, 2015, 349, 1095–1100.

[284] Wu J, Huang Y, Ye W, Li Y. CO$_2$ reduction: From the electrochemical to photochemical approach. Adv Sci, 2017, 4, 1700194.

[285] Liu A, Gao M, Ren X et al. Current progress in electrocatalytic carbon dioxide reduction to fuels on heterogeneous catalysts. J Mater Chem A, 2020, 8, 3541–3562.

[286] Guzmán H, Russo N, Hernández S. CO$_2$ valorisation towards alcohols by Cu-based electrocatalysts: Challenges and perspectives. Green Chem, 2021, 23, 1896–1920.

[287] Cauwenbergh R, Das S. Photochemical reduction of carbon dioxide to formic acid. Green Chem, 2021, 23, 2553–2574.

[288] Calmanti R, Selva M, Perosa A. Tandem catalysis: One-pot synthesis of cyclic organic carbonates from olefins and carbon dioxide. Green Chem, 2021, 23, 1921–1941.

[289] Thiounn T, Smith RC. Advances and approaches for chemical recycling of plastic waste. J Polym Sci, 2020, 58, 1347–1364.

[290] Zhang F, Zeng M, Yappert RD et al. Polyethylene upcycling to long-chain alkylaromatics by tandem hydrogenolysis/aromatization. Science, 2020, 370, 437–441.

[291] Balema VP, Hlova IZ, Carnahan SL et al. Depolymerization of polystyrene under ambient conditions. New J Chem, 2021, 45, 2935–2938.

[292] Urawa Y, Miyazawa M, Ozeki N, Ogura K. A novel methodology for efficient removal of residual palladium from a product of the Suzuki–Miyaura coupling with polymer-supported ethylenediamine derivatives. Org Process Res Dev, 2003, 7, 191–195.

[293] Galaffu N, Man SP, Wilkes RD, Wilson JRH. Highly functionalised sulfur-based silica scavengers for the efficient removal of palladium species from active pharmaceutical ingredients. Org Process Res Dev, 2007, 11, 406–413.

[294] Recho J, Black RJG, North C, Ward JE, Wilkes RD. Statistical DoE approach to the removal of palladium from active pharmaceutical ingredients (APIs) by functionalized silica adsorbents. Org Process Res Dev, 2014, 18, 626–635.

[295] Ötvös SB, Hatoss G, Georgiádes A et al. Continuous-flow azide–alkyne cycloadditions with an effective bimetallic catalyst and a simple scavenger system. RSC Adv, 2014, 4, 46666–46674.

[296] Seeberger PH. Scavengers in full flow. Nat Chem, 2009, 1, 258–260.

[297] Baxendale IR, Ley SV, Mansfield AC, Smith CD. Multistep synthesis using modular flow reactors: Bestmann–Ohira reagent for the formation of alkynes and triazoles. Angew Chem Int Ed, 2009, 48, 4017–4021.

[298] Snodin DJ. A primer for pharmaceutical process development chemists and analysts in relation to impurities perceived to be mutagenic or "genotoxic". Org Process Res Dev, 2020, 24, 2407–2427.

[299] Crosby DG. Pesticides as environmental mutagens. In: Fleck RA, Hollaender A, eds. Genetic toxicology: An agricultural perspective. New York, NY, USA, Plenum Press, 1982, 201–218.

[300] Booth ED, Rawlinson PJ, Fagundes PM, Leiner KA. Regulatory requirements for genotoxicity assessment of plant protection product active ingredients, impurities, and metabolites. Environ Mol Mutagen, 2017, 58, 325–344.

[301] ICH harmonised guideline: Assessment and control of DNA reactive (mutagenic) impurities in pharmaceuticals to limit potential carcinogenic risk. M7 (R1). International Council for Harmonisation of Technical Requirements for Pharmaceuticals for Human use (ICH), 2017. (Accessed August 5th, 2021, at https://database.ich.org/sites/default/files/M7_R1_Guide line.pdf).

[302] King FJ, Searle AD, Urquhart MW. Ranitidine – Investigations into the root cause for the presence of N-nitroso-N,N-dimethylamine in ranitidine hydrochloride drug substances and associated drug products. Org Process Res Dev, 2020, 24, 2915–2926.

[303] Nguyen T. A side reaction may have led to impurities found in valsartan heart drugs. Chem Eng News, 2019, 19th February.

[304] White CM. Understanding and preventing (N-nitrosodimethylamine) NDMA contamination of medications. Ann Pharmacother, 2020, 54, 611–614.

[305] Gerber C, Toelle H-G. What happened: The chemistry side of the incident with EMS contamination in viracept tablets. Toxicol Lett, 2009, 190, 248–253.

[306] Snodin DJ. Elusive impurities – evidence versus hypothesis. Technical and regulatory update on alkyl sulfonates in sulfonic acid salts. Org Process Res Dev, 2019, 23, 695–710.

[307] Quinto I, Staiano N, Martire G, Friscia GO, Signorini M, De Lorenzo F. Mutagenic epoxide impurities discovered in two new beta-adrenergic blocking agents. Toxicol Lett, 1980, 5, 109–114.

[308] Watson WP, Brooks TM, Huckle KR et al. Microbial mutagenicity studies with (Z)-1,3-dichloropropene. Chem Biol Interact, 1987, 61, 17–30.

[309] Frankel LS, McCallum KS, Collier L. Formation of bis(chloromethyl)ether from formaldehyde and hydrogen chloride. Environ Sci Technol, 1974, 8, 356–359.

[310] Kumar T, Ramya M, Srinivasan V, Xavier N. A simple and direct LC-MS method for determination of genotoxic impurity hydroxylamine in pharmaceutical compounds. J Chromatogr Sci, 2017, 55, 683–689.

[311] Elder DP, Snodin D, Teasdale A. Control and analysis of hydrazine, hydrazides and hydrazones – Genotoxic impurities in active pharmaceutical ingredients (APIs) and drug products. J Pharm Biomed Anal, 2011, 54, 900–910.

[312] Kedia SB, Mitchell MB. Reaction progress analysis: Powerful tool for understanding Suzuki–Miyaura reaction and control of polychlorobiphenyl impurity. Org Process Res Dev, 2009, 13, 420–428.

[313] Baczyński E, Piwońska A, Fijałek Z. Determination of 2,6-dimethylaniline and *o*-toluidine impurities in preparations for local anaesthesia by the HPLC method with amperometric detection. Acta Pol Pharm, 2002, 59, 333–339.

[314] Nessa F, Khan SA, Abu Shawish KYI. Lead, cadmium and nickel contents of some medicinal agents. Indian J Pharm Sci, 2016, 78, 111–119.

[315] Ashby J, Tennant RW. Definitive relationships among chemical structure, carcinogenicity and mutagenicity for 301 chemicals tested by the U.S. NTP. Mutat Res, 1991, 257, 229–306.

[316] Yang H, Li J, Wu Z, Li W, Liu G, Tang Y. Evaluation of different methods for identification of structural alerts using chemical Ames mutagenicity data set as a benchmark. Chem Res Toxicol, 2017, 30, 1355–1364.

[317] Luechtefeld T, Hartung T. Computational approaches to chemical hazard assessment. ALTEX – Altern Anim Ex, 2017, 34, 459–478.

[318] Teasdale A, Elder D, Chang S-J et al. Risk assessment of genotoxic impurities in new chemical entities: Strategies to demonstrate control. Org Process Res Dev, 2013, 17, 221–230.

[319] Borths CJ, Argentine MD, Donaubauer J et al. Control of mutagenic impurities: Survey of pharmaceutical company practices and a proposed framework for industry alignment. Org Process Res Dev, 2021, 25, 831–837.

Index

https://doi.org/10.1515/9783110717877-010